广州建筑图册
广州老建筑图册

OLD BUILDING

一城岁月歌谣

广州市规划和自然资源局 编

SPM 南方传媒 | 花城出版社

中国·广州

图书在版编目（CIP）数据

广州建筑图册. 广州老建筑图册 / 广州市规划和自然资源局编. -- 广州：花城出版社，2023.12
ISBN 978-7-5749-0098-1

Ⅰ. ①广… Ⅱ. ①广… Ⅲ. ①建筑文化－广州－图集 Ⅳ. ①TU-092

中国国家版本馆CIP数据核字(2023)第254875号

主编单位：广州市规划和自然资源局
承编单位：广州市城市规划设计有限公司
　　　　　广州市设计院集团有限公司

出 版 人：张　懿
责任编辑：陈诗泳
责任校对：梁秋华
技术编辑：林佳莹
装帧设计：广州市耳文广告有限责任公司

书　　名	广州建筑图册·广州老建筑图册
	GUANGZHOU JIANZHU TUCE · GUANGZHOU LAO JIANZHU TUCE
出版发行	花城出版社
	（广州市环市东路水荫路11号）
经　　销	全国新华书店
印　　刷	佛山市迎高彩印有限公司
	（佛山市顺德区陈村镇广隆工业区兴业七路9号）
开　　本	787毫米×1092毫米　16开
印　　张	12.25　1插页
字　　数	120,000字
版　　次	2023年12月第1版　2023年12月第1次印刷
定　　价	288.00元（全三册）

如发现印装质量问题，请直接与印刷厂联系调换。
购书热线：020 - 37604658　37602954
花城出版社网站　http://www.fcph.com.cn

广州的建筑，
是东西交流、南北对话、新旧共融的集大成者

南越国宫署遗址
Site of Nanyue Kingdom Palace
西汉

怀圣寺光塔
Minaret of Huaisheng Mosque
唐代

大佛寺
Dafo Temple
南汉—清

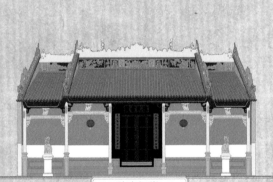
陈家祠堂
Chen Clan Ancestral Hall
清代

广州圣心大教堂
Sacred Heart Cathedral
清代

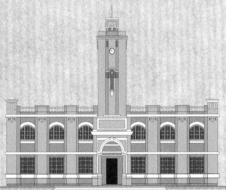
广州鲁迅纪念馆
Guangzhou Luxun Memorial Hall
1904年

江边的风景

广州大元帅府旧址
The Memorial Museum of Generalissimo Sun Yat-Sen's Mansion
清代

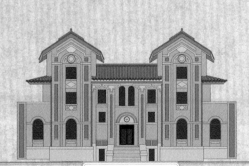
康乐园早期建筑群
Early Building Cluster of Kangle Garden
近代

中山大学孙逸仙纪念医院
Sun Yat-sen Memorial Hospital
1835年

| 光孝寺
Guangxiao Buddhist Temple
宋代 | 六榕寺
Temple of Six Banyan Trees
宋代 | 岭南第一楼
The No. 1 City Gate Tower of Lingnan
明代 | 镇海楼
Zhenhai Tower
明代 | 莲花塔
Lotus Pagoda
明代 | 西关大屋
Xiguan Mansion
清代 |

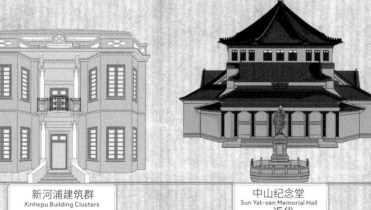
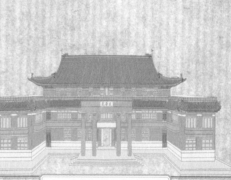

| 广东咨议局旧址
Site of Former Guangdong Provincial Consultative Council
1909年 | 文明路骑楼
Arcade Buildings in Wenming Lu
20世纪初 | 新河浦建筑群
Xinhepu Building Clusters
1922年 | 中山纪念堂
Sun Yat-sen Memorial Hall
近代 | 中山大学石牌旧址建筑群
Building Complex of Former National Sun Yat-sen University (Shipai Campus)
近代 | 兰圃
Lanpu (Orchid Garden)
20世纪50年代 |

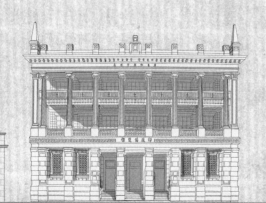

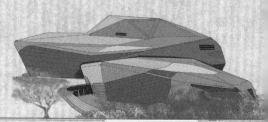

| 广东邮务管理局旧址
Site of Former Guangdong Post Administration
1916年 | 爱群大厦
Oi Kwan Hotel
1937年 | 广州白天鹅宾馆
White Swan Hotel
1983年 | 红专厂
Redtory
2009年 | 广州塔
Canton Tower
2009年 | 广州大剧院
Guangzhou Opera House
2010年 |

赤岗塔

年代 明代
地址 海珠区琶洲新港东路

写在前面
The Editor's Notes

序章 Preface

两千余年的广州城，遗存无数承载时光烙印的美好建筑，本书收录的建筑，无一不是带着岁月的沉淀，以其独有的符号，传承着广州的基因，为人们讲述这座城的故事，或风云变幻，或细水长流。

这些建筑若单纯以建造年份或所在区域划分，则历史长远，其中一些始终难有定论，而且在各区域的分布密度悬殊，无法更好领略城的全局之美。

让我们以建筑的功能划分为面，以时间轴为线，一来筛选建筑时能有所侧重，也能更好地了解建筑的由来；二来按时间顺序，也便于彼此联系及对照。

本书共分七章，《序章》是概论，从广州建筑历史的时间轴、匠人的取材智慧和建造装饰工艺三个维度开门见山，希望读者能按图索骥轻松读懂广州的建筑瑰宝；《风云》收录了帝王宫苑、城楼城墙、革命纪念地，一个个历史拐点就发生在这些建筑内，风云人物在此书写传奇；《庙宇》涵盖的建筑，跨越不同宗教，融会东西南北，从另一个侧面反映了广州的包容；《余芬》将广府人重视教育的一面显露无遗，学宫、学院、图书馆、古树成荫、桃李芬芳；《商都》选录的建筑凸显了广州的海洋性，是商贸繁华的实物见证；《人居》着力刻画广州人的人居智慧与审美；《新风》则从现代岭南建筑，看到传承与发展。

一头脊兽、一排瓦当、一列斗拱、一扇窗花、一组阶砖，都唱着光阴的歌。

The book is divided into seven chapters. *The Prologue* is an introduction to the book that helps readers easily search for historical buildings that they are interested in and better understand them. *Ride the Whirlwind* includes the sites of imperial palaces and city walls, as well as revolutionary sites. *Temples* is a collection of buildings of different religions. *Helping the Posterity* showcases the tradition of the local Cantonese in vigorously promoting educational development. *City of Commerce* presents buildings that have witnessed the flourishing business and trade in Guangzhou. *Houses* focuses on the wisdom and aesthetics of the traditional houses in Guangzhou. *New Trends* shows the inheritance and development of the traditional Lingnan architecture in the design of modern buildings.

目录
Directory

序章
Prologue

这些老房子，是凝固的时间轴
Frozen Timeline — 9

一砖一瓦，是匠人的智慧
Wisdom of Builders — 11

这些符号，
是广州的建筑文化基因
Genes of Architecture Art in Guangzhou — 13

第一章：风云
Chapter 1: Ride the Whirlwind

南越国宫署遗址
Museum of the Western Han Dynasty Mausoleum of the Nanyue King — 17

药洲遗址
Yaozhou Ruins — 19

镇海楼 | 广州明城墙
Zhenhai Tower / Ming City Wall — 23

广州大元帅府旧址
Memorial Museum of Generalissimo Sun Yat-Sen's Mansion — 27

广东咨议局旧址
Site of Former Guangdong Provincial Consultative Council — 29

黄花岗七十二烈士墓
Huanghuagang 72 Martyrs Cemetery — 33

中山纪念堂
Historical Sites of Sun Yat-sen Memorial Hall — 41

广州市政府合署大楼
Guangzhou City Hall Building — 43

十九路军淞沪抗日阵亡将士坟园
Memorial Cemetery for Officers and Soldiers of 19th Route Army Who Died in Songhu Anti-Japanese Campaign — 45

广州沙面建筑群
Historical Architectures in Shamian — 49

第二章：庙宇
Chapter 2: Temples

光孝寺
Guangxiao Buddhist Temple — 57

六榕寺塔
Temple of Six Banyan Trees — 59

大佛寺
Dafo Temple — 63

怀圣寺光塔
Minaret of Huaisheng Mosque — 67

五仙观
Temple of the Five Immortals — 71

岭南第一楼
The No. 1 City Gate Tower of Lingnan — 73

纯阳观
Chunyang Daoist Temple — 75

广州圣心大教堂
Sacred Heart Cathedral — 77

广裕祠
Guangyu Ancestral Temple — 85

留耕堂
Liugeng Hall — 91

陈家祠堂
Chen Clan Ancestral Hall — 97

序章 | Preface

第三章：余芬	第四章：商都	第五章：人居	第六章：新风
Chapter 3: Helping the Posterity	Chapter 4: City of Commerce	Chapter 5: Houses	Chapter 6: New Trends

第三章：余芬 / Chapter 3: Helping the Posterity

玉岩书院
Yuyan Academy ... 101

番禺学宫
Panyu Academy ... 103

广州农民运动讲习所旧址
Former Site of Guangzhou Peasant Movement Institute ... 105

中山大学石牌旧址建筑群
Building Complex of Former National Sun Yat-sen University (Shipai Campus) ... 111

康乐园早期建筑群
Early Building Cluster of Kangle Garden ... 121

黄埔军校旧址
Former Site of Huangpu Military Academy ... 125

广州鲁迅纪念馆
Guangzhou Luxun Memorial Hall ... 127

广州市立中山图书馆旧址
Site of Sun Yat-Sen Library of Guangzhou City ... 129

明心书院
Mingxin Academy ... 133

第四章：商都 / Chapter 4: City of Commerce

清真先贤古墓
Tomb of Saad ibn Abi Waqqas ... 137

千年古道遗址
Site of Millennial Ancient Road ... 141

琶洲塔 | 莲花塔 | 赤岗塔
Pazhou Pagoda / Lotus Pagoda / Chigang Pagoda ... 143

锦纶会馆
Jinlun Guild Hall ... 147

粤海关旧址
Site of Former Guangdong Customs ... 149

第五章：人居 / Chapter 5: Houses

西关大屋建筑群
Xiguan Mansions ... 153

耀华大街
Yaohua Dajie ... 155

恩宁路骑楼
Arcade Buildings in En'ning Lu ... 157

潘家祠街区
The Block of Former Pan Clan Ancestral Hall ... 159

文明路骑楼
Arcade Buildings in Wenming Lu ... 161

小蓬仙馆
Small Pengxian Hall ... 163

新河浦建筑群
Xinhepu Building Clusters ... 165

华侨新村
Garden Residences for Returned Overseas Chinese ... 169

第六章：新风 / Chapter 6: New Trends

山庄旅舍
Mountain Villa Hotel ... 175

兰圃
Lanpu (Orchid Garden) ... 177

北园酒家
Beiyuan Restaurant ... 179

白云宾馆
Baiyun Hotel ... 181

流花湖公园红桥、流花西苑 | 东山湖公园九曲桥
Red Bridge in the Liuhua Park / Liuhua West Garden / Zigzag Bridge in Dongshan Lake Park ... 183

5

序章
Prologue

广州建城有源可溯的历史长达2200余年,在这座传奇的老城里,留存着无数的优秀历史建筑,它们是历史长河留给我们的瑰宝,珍爱它们,读懂它们,便是读懂了我们热爱的这座城。

With a long-standing history that can be dated back to more than 2,200 years ago, Guangzhou is home to a profound heritage of historical buildings. To understand these buildings means to understand the city we love.

光孝寺 东晋时肇建的光孝寺，虽经历代重修，大雄宝殿仍保持着唐宋时的风貌

六榕寺塔 六榕塔顶，至今立有元代的千佛铜柱，印证了六榕寺的悠远历史

岭南第一楼 明初的岭南第一楼，从葱茏秀木间仰望，仍是秀丽端庄的面貌

陈家祠堂 晚清的陈家祠堂，是岭南建筑艺术成就的集大成者

中山纪念堂 20世纪30年代建成的中山纪念堂，是融汇东西的经典案例

山庄旅舍 新中国成立后，岭南建筑派绽放异彩，创作出很多优秀的作品

这些老房子，
是凝固的时间轴
Frozen Timeline

　　两千多年未移城址的广州，光阴的脉络清晰，南越王宫的地砖、怀圣寺的光塔、光孝寺的铁塔，有中原的影响，也有海上丝绸之路交融的痕迹。

　　至宋朝随着大批士族迁入，街巷布局、建筑形制、装饰纹样……各派思潮、各路审美渗入融合，在广州焕发出勃勃生机。宋人好风雅，开始在广州评出八景，江河湖海、日月晴雨、山林屋宇，无处不可入画。

　　自明代便立于越秀山山脊的镇海楼和城墙，可见城郭日渐壮大；从旧日临江而筑的五仙观和岭南第一楼，可知陆地日渐前进。

　　经明清两朝，广州成为不可或缺的商港城市，有了南来北往、东西交会的热闹与精彩，镬耳山墙、青砖石脚、三雕两塑的广府建筑，从施建到审美，已自成体系。而外廊拱券、红砖线脚、穹顶尖塔的西式建筑，亦于一口通商的清代，在广州城成片落地，千姿妙曼、各自精彩。

　　20世纪始，骑楼、洋房建筑兴起，不少在西方完成现代教育、接风气之先的岭南建筑师，相继开展糅东西所长又因地制宜的探索，岭南派建筑新风潮，自此薪火相传，历久弥新。

Guangzhou, remaining where it was for two thousand years, has been influenced by both the culture of the Central Plains and the Maritime Silk Road. The city has been embracing various schools of thoughts and aesthetic tastes since the Song Dynasty and its boundaries and land area have constantly expanded since the Ming Dynasty. After the Ming and Qing dynasties, Guangzhou established itself as a strategic business hub and trade port. As a crossroad where the north intersects with the south and the east meets the west, Guangzhou has developed a distinct Cantonese architectural style. However, western-style buildings have also been imple mented extensively, presenting a varied yet harmonious cityscape. Since the beginning of the 20th century, the Lingnan-style architecture emerged as a new trend and since then has been passed down and renewed from generation to generation.

蚝壳墙 广州水系繁多，水产丰富，乡民常用蚝壳建屋

红砂岩柱础 从明代到清初，广州人就地取材，常以遍地可见的红砂岩做基础材料

鸭屎石栏板 在更结实的花岗岩未被普及采用之前，鸭屎石和红砂岩一样，都是广州人常用的建材

花岗岩墙柱 从清代开始，更坚实、不易风化的大理石、花岗岩替代红砂岩成为建筑的主要材料

花阶砖 1907年开建的大元帅府，瓷砖是德国进口，逾百年仍完好

混凝土斗拱 进入近代，西方工业化生产的先进技术，在普及运用的时候也融入了本土的审美

一砖一瓦，
是匠人的智慧
Wisdom of Builders

珠江水系或洋洋或涓涓穿过广州城，水在广州人的生命中，至关重要。

浮居水上的疍民或以船为屋，或搭寮定居，棚寮用原木、竹子、茅草、树皮等材料筑成，简素质朴，因地制宜。

依水而居的广州人，亦善于用蚝壳砌墙，今天，小洲村、大岭村、沙湾镇等村镇，都遗有蚝壳墙老屋，既结实又美观。

自清开始，容易风化的红砂岩和鸭屎石柱基、地基、栏板，开始换成了更加适应雨水天气、结实耐用的花岗岩。各种西方现代化的建材，也开始大量被采用，德国的钢筋、法国的铁艺、意大利的石砖，一船船抵埠。

中国工匠也积极向西方学习，工业制造能力也快速成长，到了建造中山纪念堂的时候，许多建材已经国产化，从设计到制造再到施工，中国匠人与时俱进。

With the Pearl River running through Guangzhou, water plays a significant role in the lives of the local people. Living by the riverside, they built their houses with all kinds of materials, from shed and oyster shell house, red sandstone and granite, to German steel bars, French iron art and Italian stone brick etc., integrating them into the local conditions.

④

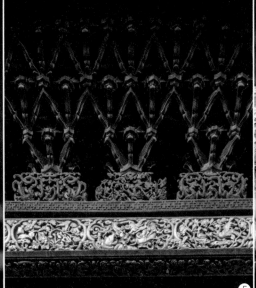
⑤

⑥

这些符号，
是广州的建筑文化基因
Genes of Architecture Art in Guangzhou

建筑是各种艺术门类的集大成者，其形制、构件、工艺等都直接反映出一座城市的独有基因，当在广州看到一座美好的老房子，你知道怎么样欣赏它了吗？

Architecture is an integration of all kinds of art. The building form, structure, component and construction technique directly reflect the unique genes of a city. However when you encounter a beautiful old house somewhere in Guangzhou, do you know how to appreciate it?

1.灰塑／ 2.砖雕／ 3.陶塑／ 4.雀替／ 5.斗拱／ 6.瓦当滴水
7.镬耳山墙／ 8.满洲窗／ 9.趟栊门／ 10.庭院
11.木雕／ 12.石雕／ 13.铁艺／ 14.八角洋房／ 15.琉璃窗花／ 16.花窗

⑩

⑭

⑮

⑯

第一章：风云
Chapter 1: Ride the Whirlwind

在充满机遇和挑战的广州，一批又一批风云人物在一座座老房子里，扼住命运的咽喉，奋笔疾书出一段段辉煌的传奇，这些老房子里，留下了悠远的历史足音。

In Guangzhou, a city of opportunities and challenges, the historical sites have witnessed the old glory of the generations of legendary figures in history.

Museum of the Western Han Dynasty Mausoleum of the Nanyue King

Year Built: The Western Han Dynasty
Add: 316, Zhongshan 4 Lu, Yuexiu District

2,000 years ago, the Palace once had a meandering stone canal which could thereby serve the guests by floating the wine cups on top for drinking. Here one may discover the remains of twelve dynasties. It is the earliest ancient garden ever discovered in China so far and the birthplace of the Lingnan Gardens.

两千年前，
广州人这样演绎曲水流觞

南越国宫署遗址

这里存留着广州十二朝旧痕，是现存年代最早的园林遗迹，岭南园林便源于此。

年代 西汉
地址 越秀区中山四路 316 号

宫署遗址内挖掘出秦代至近代的历代遗迹遗物，表明这里不仅是南越国、南汉国的王宫所在地，也是历代郡县州府的官衙所在地，是广州2200多年城市发展的历史见证。

在宫署的地下发现全石构曲流石渠，长150米，是一处人工园林水景。石渠迂回曲折，东西展开，渠道底部密铺黑色卵石。遗址共有7个时代的重叠文化层，发现83口由南越国至近代的水井，反映了不同时期的建筑文化特色。南越国宫署遗址的发掘，对研究汉代宫署园林提供了弥足珍贵的实例。

南越国御苑是目前中国乃至世界现存年代最早的园林遗迹，是岭南园林的起源之作。

1	2		5	
3	4		6	7

1. 自南越到南汉、到明清，各代的水井形制，都能在这里找到。
2—3. 秦代船坞、历代宫署、商贸重地，这里看点众多，宝物满地。
4. 两千多年未移城址的广州，在此处保有十二朝旧痕。
5. 南汉宫殿遗址上壮观的柱基布列。
6. 全石构曲流石渠原样复原。
7. 全石构曲流石渠原址。

Chapter 1:Ride the whirlwind

Yaozhou Ruins
(Site of the Islet of Immortality Pills)

Year Built: Nanhan Dynasty
Add: 86, Jiaoyu Lu, Yuexiu District

It was the royal garden of Nanhan King and its ambience was mystified by 9 pieces of gardening rocks with inscriptions of famous calligraphers or scholars, such as Mi Fu (a prominent painter and calligrapher in eleventh-century China) and Su Shi (a writer, poet, artist, calligrapher and statesman of the Song Dynasty, and one of China's greatest poets and essayists). It has remained one of the sacred places for Chinese literati.

1 2

1. 园内奇石千姿百态。
2. 历代墨客,把最大的赞美留给了药洲。

不见长生不老药,
但见文人墨客"打卡"点

药洲遗址

这处南汉王的御花园,很长的时间里,都是文人骚客的朝圣地。

年代 南汉
地址 越秀区教育路 86 号

一千多年前,南汉王朝建都广州,开国皇帝刘䶮选址现今西湖路一带建御花园,理水、置石、种药,聚方士炼丹,园内歌台舞榭,笙歌不绝。

园内鼎鼎大名的"九曜石",瘦、透、皱,形状大小色泽各异,引米芾、苏轼等文人墨客纷纷慕名前来泛舟避暑,立碑刻石,留题赋诗,将药洲捧成"网红打卡"景点。至明代,这里以"药洲春晓"列为羊城八景之一。

当年的御花园,如今仅剩药洲遗址,参天古榕树荫斑驳,奇石引人入胜,这座中国现存最早的古代皇家园林地面遗迹及园内的奇石,仍续讲着千年往事。

明月山海间，
看透广州的光阴脉络

Zhenhai Tower

镇海楼

第一章 | 风云
Chapter 1:Ride the whirlwind

明月山海间
看透广州的光阴脉络

镇海楼 | 广州明城墙

从明代开始，镇海楼便是广州城的地标，是羊城不可不看的风景，至今仍是。

年代 明代
地址 越秀区解放北路 988 号越秀公园

Zhenhai Tower

Year Built: Ming Dynasty
Add: Inside Yuexiu Park, 988, Jiefang Beilu, Yuexiu District

Zhenhai Tower has been the landmark of Guangzhou city since the Ming Dynasty. The Tower enjoys a panorama of the whole city and it has remained a view that cannot be missed even today.

1. 明代出于城防考虑建筑的镇海楼，已成为读懂广州的重要章节。
2. 屹立越秀山上的镇海楼，当年能眺望珠江浩渺的壮丽景象。

镇海楼又名望海楼、五层楼，明初兴建，历尽烽火，至今仍是广州地标，是羊城不可不看的风景。从1950年开始，镇海楼便作为广州博物馆展示广州的历史脉络，拾阶缓缓上，一件件文物看过来，读懂的，便是广州的浩瀚历史。

康有为"袖里纳纳乾坤易，眼底茫茫星汉浮"的气吞山河，和屈大均"其瑰丽雄特，虽黄鹤、岳阳莫能过之"的气象万千，便是从镇海楼的顶楼眺望，眼前正是这等山河壮阔。

第一章 | 风云
Chapter 1:Ride the whirlwind

　　明城墙，源于明代广州扩城，基于城防考虑，依旧时城北天然的山岗屏障所筑。

　　作为广州城最高处的防御工事，明城墙历经战火洗礼，至清亡，广州大面积扩城，城墙大多被拆除，仅存越秀公园的1000多米城墙。墙基多为易风化、材质疏松的明代砂岩条石，这正是明代广州城内普遍采用的建材，质朴古雅。

　　今天，明城墙一带，树木葳蕤、小径清幽，春有满目青翠，夏有凤凰似火。天长日久，条石青苔蔓生，榕树虬根盘错将城墙紧紧抱住，每一砖每一步，都是漫漫古粤时光。

Ming City Wall is the monument of the city expansion in Ming Dynasty. The over-1,000-meter foundation of the City Wall, with moss on stones and tangled banyan tree roots, tells the story of time.

1	2
	3

1—3. 明城墙也是广州扩城的印迹，硝烟已退，榕荫正浓。

25

珠水潮起潮落，帅府风起云涌

广州大元帅府旧址

这座中国最早期的水泥厂，异国情调与岭南风韵并存，孙中山先生在此度过了很长的岁月。

年代 近代
地址 海珠区纺织路东沙街 18 号

广州大元帅府旧址建于清末，原为"广东河南士敏土厂"（水泥厂）办公楼，由澳大利亚建筑师帕内设计。动荡岁月里，孙中山先生选此作为大元帅府，与众多风云人物一起在此办公，做出过许多重大决策。

大元帅府是典型殖民地外廊式建筑，以砖、木、石、钢、混凝土建成，每一层四周围以券廊，线脚装饰丰富。花瓶式护栏、竹节式排水管、百叶门窗，既善用西方建筑思潮，又贴合本地气候状况，蕴藏着浓浓的岭南风韵，实用又别致。

如今此处是国家级博物馆，展示着孙中山先生的革命轨迹，诉说着当年的风起云涌。

Memorial Museum of Generalissimo Sun Yat-Sen's Mansion

Year Built: In modern times
Add: 18, Dongsha Jie, Fangzhi Lu, Haizhu District

As part of the earliest cement plant in China, the mansion building in a combined European and Lingnan architectural styles has witnessed the life of Dr. Sun Yat-sen for many years.

1. 大元帅府的异国情调的外廊结构。
2. 这个中国最早期的水泥厂，在近代中国史上举足轻重。
3. 德国进口的地砖逾百年仍完好如新。

第一章 | 风云
Chapter 1:Ride the whirlwind

鸟语花香中，
这静穆的楼闪烁时代的光

广东咨议局旧址

这融合东西建筑风格的建筑群，是中国近现代史上的重要遗址。它见证了广东的光复，也见证了西式议会建筑与中式园林之交融。

年代 1909 年
地址 越秀区中山三路广州起义烈士陵园内

　　广东咨议局旧址隐于烈士陵园内，坐北朝南，整体配置颇具气势，既有罗马式圆形建筑风格，又不乏中国园林建筑的小桥流水，是一组中西合璧的建筑群。

　　自南至北，旧时的中式大门已改建为罗马式的4条大圆柱，石砌荷池与拱桥仍在，主体建筑为前圆后方的两层砖木结构楼，正是西方古罗马式议会建筑之形式。走进大厅，屋顶为半球形，8柱环列，空间开阔，两层内外有回廊，气势轩昂。

　　此处是中国近现代史上的重要遗址，孙中山曾九临此处议决重大国务问题，并在此宣誓就任非常大总统。毛泽东担任国民党中央宣传部代部长时，便是在这里的二楼办公。

　　作为烈士陵园公园的一部分，咨议局被鸟语花香环抱，踏进内里却静穆肃然，常年展出的展品、文献，无不折射出近代广州城乃至中国的宏大变革，帮后人读懂中国近代史、聆听历史足音。

Site of Former Guangdong Provincial Consultative Council

Year Built: 1909
Add: Inside Guangzhou Uprising Martyrs' Cemetery, Zhongshan 3 Lu, Yuexiu District

The building complex, a fusion of the Eastern and Western architectural styles, stood as a testament to the independence of Guangdong from the Qing Government, and the blending of the Western-style Council architecture and Chinese-style gardens.

1	2	4	5
3			

1—3.逾百岁的广东咨议局,在青葱树木环抱中,自有一派端庄气象。
4.广东各界代表在此开会、宣布广东脱离清政府独立,并成立广东军政府。国共合作的许多重要政策都在此制定,推动革命的不少重要会议也在这里举行。
5.1958年起,这里成为广东革命历史博物馆所在地,是通读广东近代史必去之所。

In a garden with birds chirping and flowers scenting, the building complex stands with tranquility and sparkles with the past glories. From 1958 until now, it has been the home to the Guangdong Museum of Revolutionary History and a must-visit destination to understand the modern history of Guangdong Province.

第一章 | 风云
Chapter 1:Ride the whirlwind

Huanghuagang 72 Martyrs Cemetery

Year Built: In modern times
Add: 79, Xianlie Zhonglu, Yuexiu District

The Cemetery compound is sacred with quiet remembrance and reflection. The history of the Guangzhou Uprising has been inscribed on the granite along with the names of martyrs. The park enjoys an attractive environment with towering trees and blooming flowers.

凛凛烈士碑，巍巍纪功坊
神圣与静默之美

黄花岗七十二烈士墓

年代 近代
地址 越秀区先烈中路 79 号

碧血黄花的那段血与火的历史与烈士英名一同深刻在花岗岩上，静默的墓园，连空气都凝固了。

1	2
	3

1.墓园庄严，气象肃穆。
2.陵园内树木参天，鲜花长放。
3.吕彦直的同学杨锡宗，为广州留下了不少端庄大气的公共建筑。

　　从黄花岗七十二烈士墓正门拾级而上，仰望"浩气长存"金色大字，跨过三拱凯旋门大牌坊，孙中山先生亲笔题写的"浩气长存"墓坊，登上纪功坊抚摸巨石，怀古鞠躬，肃敬之情油然而生。

　　一个世纪前，一场壮怀激烈的起义，一群青年志士以鲜活的生命为代价，打响了辛亥革命的前奏，烈士遗骸安葬于黄花岗。黄土一抔，四顾茫茫，幸岁月未忘，墓冢、碑亭、纪功坊、默池先后建造，七十二烈士墓落成。

　　墓园早期为杨锡宗设计，园内茂林修竹、黄花处处，"黄花皓月"曾被选为羊城新八景。纪功坊背后是广州现存最大的碑刻《广州辛亥三月二十九日革命记》，斑驳而威凛，清晰记录着一个个忠肝义胆的名字和他们的故事。

　　若在此地重读烈士林觉民的《与妻书》，缠绵悱恻之外必定多了一层腥风血雨。此时的墓园，静默无声，却胜有声。

以伟大之建筑，做永久之纪念
Historical Sites of Sun Yat-sen Memorial Hall

中山纪念堂

以伟大之建筑，做永久之纪念
Historical Sites of Sun Yat-sen Memorial Hall
中山纪念堂

整个大厅，跨度近50米，内无一柱，可同时容纳5000人……空间之雄伟、设计之巧妙，至今仍是教科书式的经典之作。

With a span of nearly 50 meters, the hall is column free and can accommodate an audience of up to 5,000 at the same time. This design is still considered innovative and classic even today.

Chapter 1:Ride the whirlwind

Historical Sites of Sun Yat-sen Memorial Hall

Year Built: In modern times
Add: 259, Dongfeng Zhonglu, Yuexiu District

Sun Yat-sen Memorial Hall (Guangzhou) is built in memory of Dr. Sun Yat-sen.
Standing against the backdrop of Yuexiu Mt. the Hall features a harmonious and natural symmetrical composition along the central axis. Designed by Lu Yanzhi, a famous Chinese architect, the octagonal building is a typical example of the integration of western supporting structures and detailing into a traditional Chinese mausoleum design.

以伟大之建筑做永久之纪念

中山纪念堂

从天空俯瞰，中山纪念堂像一把瑰丽宝伞，孙中山先生生前最爱的蓝色琉璃瓦，似乎永远不会褪色。

年代 近代
地址 越秀区东风中路 259 号

1. 传统中轴线以中山纪念堂为基点，规划得端庄大方，多年过去，仍觉气象万千。
2. 用水泥钢筋的西方建筑语言来重新表达东方神韵，这一点年轻的吕彦直做得非常出色。
3. 重檐卷棚歇山屋顶，华美凝重，而配色亦富丽端正。

广州中山纪念堂是中山先生逝世后为纪念孙中山先生而建，由吕彦直设计，依托越秀山，结合周边地势，前堂后碑中轴对称，和谐自然。

中式为体，西式为用，中山纪念堂做出了表范。看建筑外观，上部为单檐八角攒尖屋顶，下部在正东、南、西、北四出抱厦，重檐歇山顶；再看内部结构，采用西方先进的钢筋混凝土框架结构、剪力墙结构、钢桁架结构等建筑技术，突破了大空间建筑受中国传统木结构的限制，堂顶跨度近50米，能容纳5000人，开创了国内大空间厅堂建筑的首范，为近代公共建筑之最。

纪念堂色彩瑰丽堂皇，蓝色琉璃瓦顶、红柱、黄墙，而堂内外建筑细部装饰着的彩绘图案、雀替等，又将浓郁的传统风格藏在一处处细节中。近百年过去，仍不负"伟大建筑"的美名，足以做恒久之纪念。

1	
2	3

1. 门前月台及石阶两侧栏杆望柱头均刻有云纹、松鹤图案,和谐古雅,这月台正是1949年11月中国人民解放军进城检阅台的旧址;大门台阶前立有一对石狮,是清代广东巡抚部院门前的旧物。将城府心脏辟作市民休闲绿园,1926年修建的音乐亭上,"与众乐乐"四字见证着广州当年的"开放"。
2. 人民公园由杨锡宗设计,意式庭园布局,方形几何罗马对称形式。孙中山曾多次在此向群众演讲,宣扬民主革命。
3. 人民公园南侧有清初雕制的汉白玉石狮、石鼓,300多岁。

在广州之心，
有一座用了 80 年的市政府大楼

广州市政府合署大楼 /

年代 20 世纪 30 年代
地址 越秀区府前路 1 号

城之中心，脉之所在，与民同乐，风水天成。它是中国早期现代公共建筑的代表，也是中国近代城市市政现代化的见证。

Guangzhou City Hall Building

Year Built: 1930's
Add: 1, Fuqian Lu, Yuexiu District

Standing in the heart of Guangzhou city, this city hall building has been in service for 80 years. It is an epitome of the early Chinese modern public buildings and a testament to the modern urban construction in China.

沿着传统中轴线上的广州原点，由南往北，穿过人民公园走到尽头，便是位于府前路的广州市政府合署大楼，再往后便是中山纪念堂、越秀山。

坐北向南，钢筋混凝土结构，外观三层内分五层，飞檐翘角，黄琉璃瓦绿脊，红柱黄墙白花岗石基座，历经80年，广州市政府合署大楼至今还在"上班"。它是中国早期现代公共建筑代表之一，也是中国近代城市市政现代化的见证。

这一带自古代便是衙门官邸，是历代广州府城的"心脏"，清代先后为平南王府和广东巡抚署，1917年，孙中山倡议将此地辟为"市立第一公园"，即今天人民公园。于是，1921年，广州先于全国拥有了属于市民的公园，同年，广州市政厅成立，广州成为全国第一个建制市。

陈济棠主粤时期，看中此地交通方便，遂建市政府合署，并向社会征集设计图样，建筑师林克明的方案夺冠，其设计使合署既能连成一体、利于联络，又便于独立门户，解决拥挤之忧，大楼内各办公部门都有独立的电梯和楼梯，极为便利。

建成后一直使用至今，广州人惯有的实干、惜物，与民同乐的公仆风气，便由此可见。

Memorial Cemetery for Officers and Soldiers of 19th Route Army Who Died in Songhu Anti-Japanese Campaign

Year Built: 1933
Add: 113, Shuiyin Lu, Tianhe District

In his design, architect Yang Xizong referenced the monumental design in the West to create a grand and well-organized cemetery in elegant style.

碧血丹心犹在此，
清风拂面过松林

十九路军淞沪抗日阵亡将士坟园

当年，将士们在他乡铁血干城；如今，将领与手足在家乡共长眠，丹心与日月同辉。

年代 1933
地址 天河区水荫路113号

1	3	4
2		

1. 远远看去，高19米的圆柱体纪功碑将视线收窄，访客自然将脚步放缓。
2. 陵园内以陵园最高级别的松树簇拥墓道。
3. 吕彦直的同学杨锡宗，为广州留下了不少端庄大气的公共建筑。
4. 巍峨凯旋门记载丹心。

"国难当前，同胞猛省"，1932年"一·二八"淞沪抗战在上海打响，十九路军淞沪抗日阵亡将士坟园却于次年建于广州，皆因十九路军基本由广东籍子弟组成，总指挥蒋光鼐、副总指挥兼军长蔡廷锴也是广东人。

陵园乃当年华侨捐资建成，由杨锡宗设计，将西方大型纪念性建筑风格引入中国，规模宏伟，布置严谨，造型典雅。

南北走向的中轴线墓道长百余米，松树遍植，纪功碑、弧形柱廊、抗日阵亡将士题名碑与抗日亭等主体建筑均用花岗岩石砌就。其中先烈纪念馆原名"墓庐"，庄重肃穆，穿过正面10根古希腊式立柱，可见墙刻将领蒋光鼐与蔡廷锴的题词，将军已逝，意气不灭。而北端的纪功碑，其太阳状基座与背后12根罗马半月形廊柱相呼应，象征抗日将士与日月同辉。

夏日漫步墓园，南风拂过松树，确是避暑胜地，仰望凯旋门上"碧血丹心"四字，纵使一条路将坟园分割东西，也无法割舍人们对先烈的追思。愿英魂安宁，庇现世和平。

时间的荫，建筑师的梦工场
Historical Architecture in Shamian
广州沙面建筑群

第一章 | 风云
Chapter 1: Ride the whirlwind

时间的荫，
建筑师的梦工场

广州沙面建筑群

百岁的榕与樟，欧陆风情画般的街景，这座沙洲小岛是广州的宝。

年代 清代—20世纪30年代
地址 荔湾区沙面大街

　　沙面这座沙洲上的小岛，被珠水环抱，与城的烦嚣相隔。百年的榕与樟，浓荫匝地，入岛便见满目青葱；百年的西洋房，变化万千，漫溯更觉营造之美。它也是中国近代史与租界史的重要见证。

Historical Architecture in Shamian
Year Built: From Qing Dynasty to 1930's
Add: Shamian Dajie, Liwan District

With the picturesque European style streets lined with centurial banyan trees and camphor trees, the Shamian Island was once the dream place for architects and now the treasure of Guangzhou.

```
    2
1
```
1. 汇丰银行旧址。
2. 台湾银行旧址。

沙面百岁的老樟树吐送清芬，清晨水汽弥漫花气更觉悠长。路德堂旁长着棵老木棉，三四月间，花的红霞映着明黄的哥特尖顶，莫名的美。

沙面的发展始于第二次鸦片战争时期，当时沙面沦为英、法两国租界，英法殖民者挖涌使其与北面陆地分开，沙面从此成为一个孤立小岛，仅通过东、西两桥进出。1859年起，英法两国对沙面进行统一规划，按照西方近代城市"井"字形进行布局，岛内主次道路纵横正交，河堤、街心花园、公园和建筑等构成欧陆式街区风貌。

至20世纪40年代，沙面租界的公共设施已基本完备，主要建筑有领事馆、教堂、银行、邮局、电报局、医院、酒店和住宅等，还有俱乐部、酒吧、网球场和游泳池等设施，西方新古典式、仿哥特式、仿巴洛克式等建筑风格百家争鸣。

当时的沙面如同今天的珠江新城，也是建筑的博物馆、建筑师的梦工场。有人特意前来一看这些在沙洲淤泥上拔地而起的建筑，它们规模完整、风格多变。而同在沙洲上的榕樟老树年年葳蕤，是为后人可享的荫。

1	3
2	

1.沙面的新古典主义建筑。
2.原汇丰银行宿舍。
3.沙面外廊式建筑。

Once the concession of Britain and France, Shamian Island was built on a grid layout that can be found in many western modern cities. Roads on the Island intersect with each other orthogonally. Within a European style block, the buildings of the consulates, church, bank, post office, hospital and hotels in Neoclassical, Gothic or Baroque styles were completed one after another on the island.

第二章：庙宇
Chapter2: Temples

　　相信机遇，相信拼搏，相信北帝可保风调雨顺，相信海神可庇海不扬波，相信列祖列宗在天之灵可佑平安，广州人也相信行善修心，能得安宁与平静，从前广州人，将最大的敬畏心献给了心中的庙宇。

　　而这一座座庙宇与宗祠，就是心之憩所。

In the past, the people of Guangzhou held temples and ancestral halls in the highest regard, making them sanctuaries of their hearts.

岭南第一刹，遍地是千年宝物
Guangxiao Buddhist Temple
光孝寺

第二章 | 庙宇
Chapter 2: Temples

岭南第一刹，
遍地是千年宝物

光孝寺

光孝寺从前便是南来北往的得道高僧交会之所，上千年的宝物印证着这处佛教东来的最初福地。

年代 唐—宋
地址 越秀区光孝路109号

  广州有句老话：未有羊城，先有光孝。光孝寺的地位，自是无比尊崇，高僧们涉重洋而来，落地第一站，便是"最初福地"的光孝寺。

  光孝寺的渊源要溯至汉代，大雄宝殿后的诃子树，虽是后来补植，也是宋代之物了。同是宋代之物的，还有殿后一列石狮望柱头，比石狮更老的，是南汉时代的两座铸工精巧的铁塔。寺内的一座石经幢，更是岭南十分珍贵的唐代文物，也是光孝寺内现存有绝对年代可考的最早石刻。经幢刻有梵文和汉文大悲咒，得名大悲幢。

  而中国佛教著名的典故——风动、幡动，还是心动的偈语，也是由六祖慧能在寺里答出，他后来在寺中的菩提树下重新剃度，落发存在寺中瘗发塔内。

  达摩洗钵池、南汉铁塔、石经幢、慧能瘗发塔……动辄满目上千年的瑰宝，每一物皆有典故，光孝寺，自是福地。

Guangxiao Buddhist Temple

Year Built: From Tang to Song Dynasties
Add: 109, Guangxiao Lu, Yuexiu District

Known as the No. 1 Buddhist Temple in the Lingnan region, Guangxiao Buddhist Temple used to pool together many eminent monks. Treasures left from over a thousand years ago are evidence that the Temple was one of the blessed places where Buddhism was first preached in China.

1. 宋代的石狮望柱头。
2. 院内菩提树、诃子树、榕树古树婆娑。
3. 六祖慧能之瘗发塔。
4. 南汉时期的铁塔，精妙绝伦。

1	2
	3	4

第二章 | 庙宇
Chapter 2: Temples

当年羊城第一高，
王勃、苏轼留墨宝

六榕寺塔

六榕花塔一度是广州的制高点，古榕、舍利塔、六祖铜像，千余年间，寺庙命运浮沉，星宿雁过留声。

年代 宋代
地址 越秀区六榕路 87 号

六榕寺得名，源于大文豪苏轼被寺里生机勃勃的六株古榕震撼，题下"六榕"墨宝。

王勃在此留下的一篇《宝庄严寺舍利塔记》，字字华丽、句句工整，"仙楹架雨，若披云翼之宫；采槛临风，似遏扶摇之路"，六榕寺当年的庄严与华美，被少年才子以磅礴奇俊的文字定了格。

南汉时寺和塔皆毁于兵火，塔基内舍利完好。宋代在原址上复建为花塔，塔高57.6米，直到清代石室教堂建成之前，它一直是广州的最高建筑。花塔高耸于广州城内建筑群之上，不仅是登临胜地，还是珠江航道的重要标识。

今天的六榕寺，规模虽已减小，但宋代的六祖铜像仍在，宋时的格局犹存，塔顶的元代千佛铜柱安好。清代补种的榕树，此时正浓荫匝地，一派盎然生趣。

Temple of Six Banyan Trees

Year Built: Song Dynasty
Add: 87, Liurong Lu, Yuexiu District

Was it the banyan tree, the stupa, or the bronze statue of the Six Patriarch that inspired Wang Bo and Su Shi, poets of the Tang Dynasty, to leave their calligraphy here? Flower Pagoda in the Temple was once the highest point of Guangzhou city and an important navigation mark of the Pearl River Waterway.

The tall and beautiful Flower Pagoda is one of Guangzhou people's favorite spots to enjoy a panoramic city view. The Temple of Six Banyan Trees has always remained a sacred place among the local people.

1	2	3
	4	5

1.在石室圣心大教堂建成之前，花塔一直是广州最高的建筑。

2—5.秀美挺拔的六榕塔，又被老广亲切地叫作花塔，又美又高的花塔，是广州人最爱的登高远眺的第一胜景，而六榕寺也成为老广心目中极尊崇的丛林。

Chapter 2: Temples

人过大佛寺，
大佛看遍人间事

大佛寺/

广府五大丛林之一，仿官庙制式，兼岭南风韵，因屠城之王的救赎而修建，因岭南之冠的大佛而得名，大佛吞下千年人间烟火。

年代 南汉—清
地址 越秀区惠福东路惠新中街 21 号

Dafo Temple

Year Built: From Nanhan to Qing Dynasties
Add: 21, Huixin Zhongjie, Huifu Donglu, Yuexiu District

As one of the Top Five Buddhist Temples in Guangzhou, Dafo Temple is an imitation of the official temple style with Lingnan features. It was named so because it housed the biggest Buddha statue in Lingnan region.

1	3	4
2		

1.南汉兴建的大佛寺，虽经历次重创，今天依然在繁华闹市留下了一处香火袅绕的清静地。
2.曾经，大佛寺里的大佛，在老广心目中相当有分量。
3.吞脊兽的背后，是大佛寺新建成的雄伟藏经阁。
4.历朝重建，唯柱基长存永固。

　　大佛寺为南汉开国皇帝刘䶮上应天上二十八宿而建。历经多朝毁坏，于清初平南王尚可喜仿京师官庙制式，融合岭南地方风格，重建殿宇。

　　大雄宝殿气势磅礴，殿内安南王捐赠的巨大楠木柱名贵罕有自不必讲，正中供奉黄铜精铸的三尊三世佛像，至今仍为岭南之冠，"大佛寺"因而得名。当时粤人以回文巧句大赞"人过大佛寺，寺佛大过人"。

　　雍正年间寺院范围扩大，佛事兴旺，声名远播，成为广府五大丛林之一。如今殿内大佛金光明照，作为大型佛教图书馆，藏书22000种19万册，惠泽世人。

一江珠水新月弯，光塔照见来时路
Minaret of Huaisheng Mosque
怀圣寺光塔

1 | 2 | 3
1. 唐时兴建的光塔。
2. 怀圣寺内古朴清幽。
3. 看月楼的檐角。

Minaret of Huaisheng Mosque

Year Built: Tang Dynasty
Add: Inside Huaisheng Mosque, 56, Guangta Lu, Yuexiu District

The Mosque was the place where the first Muslims who came to Guangdong performed prayers. The Minaret, a remnant from the Tang Dynasty, was once the lighthouse for ships on the Pearl River.

一江珠水新月弯，
光塔照见来时路

怀圣寺光塔

第一批来粤的伊斯兰教徒就在这里做礼拜，光塔是唐代遗物，曾每夜竖灯引航，映照珠水新月。

年代 唐代
地址 越秀区光塔路 56 号怀圣寺院内

　　怀圣寺建于唐初，古时位于珠江边，是伊斯兰教传入我国后最早建立的清真寺之一，为纪念伊斯兰教创始人、至圣穆罕默德，故名怀圣寺，因寺内有一光塔，又称光塔寺。

　　怀圣寺于元代被焚毁，后经历次重修，唯光塔为唐代遗物，礼拜堂台基及月台为明代遗物，由砂岩制，栏板雕刻精致古朴的暗八仙及如意图案。

　　光塔由青砖砌筑，圆筒形塔身向上收窄，为国内现存伊斯兰教建筑最早最具特色的古迹之一。古时塔顶每晚高竖导航明灯，为船舶探照指路，是海上丝绸之路的重要文化遗产，见证了广州千年繁华。

羊城为什么叫羊城
五仙观里有答案
Temple of the Five Immortals
五仙观

1	2	
3	4	5

1—2. 五位仙人持五谷骑五只仙羊来帮助羊城人民的传说，这里有官方版本。
3. 明代风格的石马石人。
4. 石湾琉璃鳌鱼也是广府建筑常见的脊兽。
5. 后殿完好地保留了鲜明的明代风貌。

Temple of the Five Immortals

Year Built: Ming Dynasty
Add: 233, Huifu Xilu, Yuexiu District

The Temple offers an answer to the question why Guangzhou is nicknamed the City of Rams. The Temple presents both the official building style of Ming Dynasty, characterized by elegant and flowing outlines, and the local architectural features in northern and southern China.

羊城为什么叫羊城，五仙观里有答案

五仙观

既有明代官式建筑的隽美流畅，也有南北交融的地方特色，亲临此地重读羊城典故，有"官味"。

年代 明代
地址 越秀区惠福西路 233 号

　　传说中五位仙人骑着五色仙羊将稻穗赠予广州人，五仙观便是为祭祀此五仙而建造的谷神庙。

　　五仙观历史上多次迁址重建，现存的乃明初建造。后殿面阔三间、进深三间，绿琉璃瓦重檐歇山顶，既保留了明代早期的木构做法，又具有南方建筑特色，是广州地区现存少数明代殿堂建筑之一。

　　后殿东侧的原生红砂岩石上，有一古时珠江水冲刷形成的脚印状凹穴，称为"仙人拇迹"，被视为晋代以前"坡山古渡"遗迹，是当年珠江江岸遗迹。五仙观内还存有宋至清的碑刻十四方及石麒麟一对。

　　明、清两代先后以"穗石洞天"和"五仙霞洞"列为羊城八景之一。羊城寻根，典在此处。

1	3
2	

1.岭南第一楼又叫禁钟楼，比镇海楼建造得早，楼基用红砂石砌筑，中通往来，做城门状。
2.明初铸成的大铜钟，因象征灾祸，老广们都害怕听到它鸣响。
3.当年第一楼后便是人流如织的渡口，如今仍可见到当年遍布珠江岸的红砂岩。

The No. 1 City Gate Tower of Lingnan

Year Built: Ming Dynasty
Add: 233, Huifu Xilu, Yuexiu District

The early Ming Dynasty City Gate Tower is a well-preserved example of the elegance and solemnity of the official building style. The Tower also houses the largest and best-preserved ancient bronze bell that has been discovered in Guangdong Province so far.

一口不受喜欢的铜钟，镇着岭南六百年

岭南第一楼

这栋明初的城楼，完好地保存着官制建筑的端丽庄严，楼上是广东省现存最大最完整的古铜钟，它若安好，岁月静好。

年代 明代
地址 越秀区惠福西路 233 号

　　岭南第一楼又名禁钟楼，位于五仙观后，始建于明初，年份早于镇海楼，是广州地区唯一的明代钟楼建筑。

　　岭南第一楼分上下两层，上层为木结构建筑，下层为红砂岩基座。

　　二楼正中悬挂一口明代青铜大钟，钟底下以方形竖井直通门洞，形成一个巨大的共鸣器，这独具匠心的设计，可使钟声洪亮悠扬，据说叩之可"声闻十里"。此钟是作为遇火警非常事故时召救之用，无事禁止撞击，因名"禁钟"，是广东省现存最大最完整的古铜钟。

漱珠声里朝星斗，
凤凰木下显灵官

纯阳观

广州最大的道观，藏着广东最早的天文台，昔日蓬莱仙山之风光，引各路名人留下墨宝。

年代 清代
地址 海珠区瑞康路 268 号

  清道光年间，岭南高道李明彻在漱珠岗修建纯阳观，坐北向南、依山而建的殿宇错落有致，是广州现存最大的道教宫殿，至今仍有"北到三元宫，南去纯阳观"之说。

  穿过正门拾级而上，高敞的山门上刻有"纯阳观"三个大篆字，门头横额和对联则出自清十三行首富潘仕成之手。

  纯阳宝殿东北边有一栋四方形碉楼式建筑，便是朝斗台，这是广东最早建立的天文台，比香港皇家天文台还早建几十年，是李明彻为观测天文星象而建。昔日登台远眺，云山珠水尽收眼底，前景则是岗上的梅花与凤凰木，是不可多得的写生胜地；如今登楼远望，凤凰花红艳、摩天楼接天，漱珠岗前珠江水远退，唯香烟不停歇。

  李真人在此编写出的《寰天图说》，为中国气象观测学和地图绘测学立下了大功。史上一批接一批文人骚客，被李真人的才情和漱珠岗的清幽所吸引，如高奇峰、高剑父、陈树人等便承继"两居"师业，长年累月在纯阳观作画研究，为岭南画派奠下基础。

Chunyang Daoist Temple

Year Built: Qing Dynasty
Add: 268, Ruikang Lu, Haizhu District

Being the largest Daoist temple in Guangzhou, it houses the earliest observatory in Guangdong Province. The Temple, with its once picturesque view resembling the fabled mountains of the immortals, attracted many literati who left inscriptions of their calligraphy here.

第二章 | 庙宇
Chapter 2: Temples

广东石匠领衔的瑰丽天主堂

广州圣心大教堂

将中国元素和本土建材融入一座哥特式建筑之中，这座全石结构教堂，是匠心独运的奇迹。

年代 清代
地址 越秀区一德路旧部前 56 号

广州圣心大教堂是国内现存最宏伟的双尖塔哥特式建筑之一，也是全球四座全石结构哥特式教堂建筑之一（另外三座是巴黎圣母院、威斯敏斯特教堂、科隆大教堂），由于教堂的全部墙壁和柱子都是用花岗岩石砌造，故广州人又称之为"石室"或"石室耶稣圣心堂""石室天主教堂"。

Sacred Heart Cathedral

Year Built: Qing Dynasty
Add: 56, Yide Lu, Yuexiu District

Reputed as the Notre Dame de Paris of the Far East, the Cathedral was built by Guangdong local masons who successfully incorporated the Chinese elements and local construction techniques into the Gothic architecture.

The Cathedral stands in the bustling downtown of Guangzhou. As one of the few cathedrals in the world to be entirely built of granite, it is a miracle of ingenuity.

1
2
3

1. 石外墙装饰。
2. 用石头雕刻镂空的圆形玫瑰窗。
3. 门框外有层层退入的7对柱线与7层尖拱肋。

　　1863年12月，圣心大教堂的奠基典礼空前盛大，传教士明稽章专门从罗马和耶路撒冷运来泥土1公斤，寓意天主教创立于东方之耶路撒冷、兴起于西方之罗马。教堂东侧墙角下"JERUSALEM1863"和西侧墙角下的"ROME1863"刻字至今依然清晰可见。

　　1864年，法国教会专门请来两位建筑师Vonutrin和Humbert主持，仿巴黎的圣克洛蒂尔德设计，由潮汕石匠蔡孝任总管工，1888年教堂竣工，以当时技术，十分高效，须知巴黎圣母院建了87年，德国科隆主教堂更是建了7个世纪才完成。

　　石室平面为天主教正统形制拉丁十字式，正面一对尖顶石塔高耸入云，内部尖券拱肋高叉穹隆，透出神秘而轻巧的飞升感，向世人展示着典型哥特式教堂建筑风格。中殿外墙采用哥特式建筑特有的飞扶壁和扶垛，为空心墙、巨大彩色玻璃窗提供支撑之力。

　　教堂所有门窗都以法国制造的较深的红、黄、蓝、绿等七彩玻璃镶嵌，使室内光线终年保持柔和，形成慈祥、肃穆的宗教气氛。正立面和东西侧立面各有一个直径近7米的圆形玫瑰花窗，绚丽大气。

　　这幢奇特的哥特式建筑里，融入了不少中国元素和建造的本土做法，如用糯米、桐油代替水泥，防水、稳固又节约成本；穹顶石块也从中凿双孔用铁枝穿起来，使之更牢固；教堂楼顶的出水口，改成了中国的螭首；地板也将原先设计的石块改成广府建筑常见的大阶砖，更利于防潮。

　　教堂矗立于广州城的繁华地带，自古商贾云集，船来船往，当年外国船只一进珠江，即可见教堂。神圣与世俗，在此分割，又在此交融。

第二章 | 庙宇
Chapter 2: Temples

1	2
	3

1. 圣心大教堂内部。
2—3. 光透过玻璃留下绚丽光斑。

81

融汇南北的祠堂，
丈量岭南建筑史的标尺
Guangyu Ancestral Temple
广裕祠

Guangyu Ancestral Temple

Year Built: From Song to Qing Dynasties
Add: Qiangang Village, Taiping Town, Conghua District

It is an ancestral hall recognized as a living example of Chinese architecture spanning from the Song Dynasty to the modern day. It combines features of both southern and northern architectural styles and remains a landmark in the history of Lingnan architecture.

融会南北的祠堂，丈量岭南建筑史的标尺

广裕祠

一座祠堂，记载了宋明清至今的印迹，时光流转一屏画。

年代 宋—清
地址 从化区太平镇钱岗村

1. 广裕祠。
2—3. 精彩的斗拱。
4. 广裕祠正门。
5. 西更楼上的封檐板，被称为"岭南清明上河图"。

在广州从化太平镇，有一处古老的村落，古老护城河内的一湾碧水将它环抱，一片繁茂的荔枝林为它遮阴，就在那个被时光和世人遗忘的安静角落里，它默默地走过了宋、明、清……并坚守至今，完好地保存着800年来不灭的记忆。在人们眼中，钱岗村是空寂、古朴、深邃，进入其中可享受远离尘嚣的静默。然而，在考古界，这座明清风格的村落大名鼎鼎：它是荣膺"2003年度联合国教科文组织亚太地区文化遗产保护杰出项目奖"第一名的"广裕祠"所在地。

广裕祠堂建筑是珠三角地区祠堂中罕见的兼具南北风格的建筑。悬山顶、山墙承重、素瓦屋面以及正门前的八字形照壁，这种北方地区用来阻挡风沙的照壁，在岭南尚属孤例。而广裕祠中院的天井、石井台却有岭南建筑的风韵，讲究四水归堂，旁边的檐廊上也多有岭南的砖雕、木雕。

广裕祠建成至今，经历6次大修，于正檩、横枋底部有明确的重建时间，与每一进大式木作结构和雕刻手法、风格完全吻合，为岭南建筑中所罕见，被专家称为"岭南建筑历史的一把标尺"和"非常宝贵的建筑标本"。

瓜瓞延绵的书香门第，逾千秋亦流芳

Liugeng Hall
留耕堂

瓜瓞延绵的书香门第,逾千秋亦流芳
Liugeng Hall
留耕堂

| 1 | 2 |
| | 3 |

1. 仪门端庄大气，引人端正仪容，恭敬三分。
2. 沙湾飘色游行。
3. 沙湾的建筑颇可看，沙湾人也多是巧匠，民艺相当出名。

瓜瓞延绵的书香门第，
逾千秋亦流芳

留耕堂

祭祀先贤，考取功名，流芳后世。

年代 元一清
地址 番禺区沙湾镇

留耕堂又称何氏大宗祠，堂名来源于该祠堂的对联：阴德远从宗祖种，心田留与子孙耕。祠堂最早建于元代，历经多次重建，现祠堂为清初所建。

头门门顶的横梁，33个三重如意斗拱，雕刻内容花样百出，或奇花异卉、飞禽走兽，或历史故事人物，无不栩栩如生。整座前门，梁、枋、斗拱共同构成一组岭南建筑艺术珍品，精美夺目。

石碑坊门额是明末广东大儒陈白沙所书"诗书世泽"的石刻，褒扬沙湾何氏乃书香门第，历代考取科举功名者辈出。而碑坊额上"三凤流芳"四个苍劲大字，是为表彰沙湾何氏祖先在北宋后期考取进士的"何家三凤"三兄弟。

月台基石是一列元、明年间的古石雕，图案精巧，刀法浑朴自然、玲珑剔透，珍贵异常。

站在月台上，环顾四周，北面象贤堂和东西庑廊的60多条大柱，错落有致，虚实相间，像琴弦般，奏出悠扬婉转的旋律，留耕堂亦如遐迩闻名的沙湾音乐，乐韵千秋百世流芳。

Liugeng Hall

Year Built: From Yuan to Qing Dynasties
Add: Shawan Town, Panyu District

Liugeng Hall is also known as the Grand Ancestral Hall of He Family. It was used to worship ancestors. The He Family has produced many scholars throughout history and its lineage flourishes after several hundred years.

民艺之大成
Chen Clan Ancestral Hall
陈家祠堂

民艺之大成，时光之分割线
Chen Clan Ancestral Hall
陈家祠堂

Chen Clan Ancestral Hall

Year Built: Qing Dynasty
Add: 34, Enlong Li, Zhongshan 7 Lu, Liwan District

Many skilled craftsmen worked together to make the Ancestral Hall a perfect example of the grand layout typical for the traditional Lingnan architecture. Every detail was meticulously designed under strict rules to create a timeless masterpiece.

民艺之大成，时光之分割线

陈家祠堂

一众能工巧匠，展现了传统岭南建筑的盛大格局，且每个细节都有其设计规则，华丽精美毫不偷懒，古祠得以流芳。

年代 清代
地址 荔湾区中山七路恩龙里 34 号

```
      2
1  ———
      3
```

1. 精彩绝伦的灰塑。
2. 屋宇连亘，三雕两塑相辉映。
3. 岭南建筑里常见的砖雕，在这里被用到极致。

按理来说，一个有着两千多年历史的城市，为何总是由这一个才一百来年的建筑，在旅游景点中独占鳌头，博得中外游客的赞誉？何况，它不过是陈氏一姓之祠堂。

只因自陈家祠堂之后，广州再也难有这种布局严谨宏大、一招一式都恪守传统礼仪、宗族文化，又融合生活美学的岭南建筑出现，也再无从招纳如此众多的能工巧匠，一砖一木、一石一铁，把一个国度的历史传说、神话寓言，一个民族最朴实或者说最现实的愿望期盼、生活憧憬，这样集中地、具体而细微地雕刻展现出来，毫不吝啬地将最精美的岭南木雕、砖雕、石雕、陶塑、灰塑、彩绘画、铜铁铸的极致工艺融为一堂。

它如同时光的分割线，是古老广州步入现代城市前，对传统文化的最后一次礼数周全的深刻致礼。

第三章：余芬
Chapter 3: Helping the Posterity

广府人于教育，向来热忱，往往重才重德并不甚重功名，像鸿儒陈白沙，像志士康有为，无须功名背书，却从学者众。而现代文明的开蒙，广州人也总是开风气之先，因此，在这片热土上，各种育人育德、赠人余芬的历史建筑数不胜数，今天仍有余芬。

The Cantonese people have always been passionate about education and bold in taking the lead. This has therefore led to the establishment of numerous academic buildings in Guangzhou at different historical periods.

Yuyan Academy

Year Built: From Southern Song to Qing Dynasties

Add: Inside Luofeng Mountain, Luogang Jie, Huangpu District

Tactfully designed and placed near the mountain and by the valley, this long-standing academy serves a triple function as an academy, ancestral hall, and temple. It is worth visiting even for its time-honored litchi trees.

广州书院最早处，千年古荔伴泉生

玉岩书院

倚山傍谷，集书院、祠堂、寺庙于一身，这座"长寿"书院因着巧选址和妙设计，避过纷扰，伴着古荔幽雅地静守于萝峰。

年代 南宋—清

地址 黄埔区萝岗街萝峰山麓

| 1 | 3 | 5 |
| 2 | 4 | |

1. 玉岩书院与萝峰寺入口台阶，树影婆娑。
2. 余庆楼与玉岩堂之间是石砌方池，尽显文人志趣。
3. 萝峰寺内一棵千年古荔依旧茂盛，如书院之长盛不衰。
4. 拾级而上玉岩堂，一探这座集门、楼、廊、舍于一身的建筑。
5. 东门外还有8柱重檐歇山顶的山高水长亭，二进院落的文昌庙、"天衢云路"石牌坊以及候仙台、招隐亭和金花庙。

萝岗自古是盆地，萝峰乃其最高处。十里梅花浑似雪，自明代后，文人骚客纷至萝峰山一带赏"萝岗香雪"。时移世易，一度柑橙荔枝成游人新宠，今日香雪复种，渐成盛况。唯萝峰山间的古书院前，人烟稀少。

玉岩书院创建于南宋，是广州历史最早的书院之一，历代多次重修，所幸心远地偏，少受战乱所扰，格局尚完整。

玉岩书院建筑群极富岭南地域文化特色，倚山傍谷一字排开，上下两进，横向分东、中、西三座，结构紧凑。这种书院、祠堂、寺庙的组合布局，在岭南建筑中实属少见。建筑组群东厅西斋，檐廊相接，建筑体南低北高，利于通风采光，且因山势而错落，室内外翠色交融，奇花佳木，冬暖夏凉，清香满堂。

以余庆楼与玉岩堂为主体，大小厅堂、厢房近20间，以庭园和天井分隔、廊道相连，一收一放，明暗相间。巧妙的是，三合院中的东西厢都简化为连廊或直接融入庭院，建筑群内几乎全部空间都东南或南朝向，通风、采光、防晒，顺理成章，又见民间建筑之灵活性。余庆楼为重檐歇山顶，玉岩堂是单檐，余庆楼三面与玉岩堂正面合围，构成"七檐滴水"之庭院，若遇雨天四方泄水或晨雾瓦缘垂珠，意境难得。

山间千年古荔，伴石泉共生，后山百年糯米糍荔枝遍植，处处清幽入胜。

1	3
2	4

1—4. 作为岭南地区最重要的礼文之所，番禺学宫虽兴于明代，格局在清代却大为改动——木柱改为更耐潮的石柱，前庙后学式的格局也改为左庙右学。

红墙育孺子，
树人两清芬

番禺学宫

身处烦嚣心脏，曾是岭南最高学堂，现在葱郁古木间低调栖身，番禺学宫以其建筑与景观诠释了儒释道的合一。

年代 明—清
地址 越秀区中山四路 42 号

番禺学宫始建于明洪武三年（1370），是祭祀圣人孔子的"孔庙"，也是明清时期番禺县儒生学习的"学宫"。与广府学宫、南海学宫，并称广州三大学宫，另两处皆已湮灭，番禺学宫实为广州唯一仅存的学宫。

学宫经两次大火烧毁、多次改建，许多建筑不复存在。大成门，寓意孔子集古代文化之大成，正中书"番禺学宫"四字尚见。

当年经过石鼓进了大成门，便是居于学宫正中的宫殿——大成殿，建筑面阔五间，进深三间，歇山顶，正脊之上有石湾文如壁所造"双龙戏珠"陶塑琉璃脊，这是学宫中唯一有凸出龙形的岭南广府式彩塑横脊，也是清代广东琉璃脊饰的代表作。作为学宫的主体建筑、奉祀孔子的主殿，当年盛大的入学礼便在此举行，时移世易，学宫里浓厚的育人之气息留有余芬，千年文脉早沁入一石一木，无处不生根，无处不兴旺。

Panyu Academy

Year Built: From Ming to Qing Dynasties
Add: 42, Zhongshan 4 Lu, Yuexiu District

As the most important venue for etiquette and culture in Lingnan region, Panyu Academy is modestly nestled amid amid verdant ancient trees. Its architecture and landscape effectively illustrate the unification of Confucianism, Buddhism and Taoism.

1. 培育英才的番禺学宫，亦是点燃星火的广州农民运动讲习所。
2. 庭院里遍植木质坚硬、生命顽强的铁刀木，铁刀木秋冬开花，落一地明艳黄花。
3. 用花岗岩雕琢的棂星门。

红墙端肃古树勃发，
余芬仍在薪火未绝

广州农民运动讲习所旧址

古树红墙下有一处英雄的摇篮，青年毛泽东、周恩来执教的身影犹在。

年代 20世纪20年代
地址 越秀区中山四路42号

　　越秀老城，地铁上盖，一组红墙黄瓦、古朴庄重的建筑群尤其显眼。明清时期的番禺学宫，于20世纪20年代成为中国农民运动的摇篮——农讲所，如今建筑群门楼正中，悬挂着刻有"毛泽东同志主办农民运动讲习所旧址"的横匾，为周恩来亲笔题写。

　　1926年5月，毛泽东在此担任讲习所所长，为北伐储备人才。当年毛泽东就站在大成殿内讲课，青年学员奋笔疾书，当年风华正茂的学员，后来一一成长为南昌起义、秋收起义、广州起义等运动的骨干。

　　时光流转，这里的历史足音并未远离，当代青年也常有组织参观农讲所接受思想洗礼，或追寻先辈的足迹，或遥想昔日建筑的辉煌。那古树生机勃发，如一首岁月献给此城的赞歌。

Former Site of Guangzhou Peasant Movement Institute

Year Built: 1920's
Add: 42, Zhongshan 4 Lu, Yuexiu District

This cradle of heroes, nestled beside old trees and red walls, still evokes memories of the days young Mao Zedong and Zhou Enlai lectured here.

1	2		5	6
3	4		7	

1—2.番禺学宫的灰塑和陈家祠的同为清代名厂文如璧所造。
3—4.岁月更迭,此处香樟、铁刀木、潺槁木姜子参天,处处皆是南国秀雅。
5—7.前来参观学习的学子们四时不绝,沾不竭的灵气,汲不灭的激情。

中国人自己规划建设的山水校园
The Chinese-designed and -built landscaped campuses
中山大学石牌旧址建筑群
Building Complex of former National Sun Yat-sen University (Shipai Campus), as well as building complexes of South China University of Technology and South China Agricultural University

中山大学石牌旧址建筑群是20世纪30年代，按孙中山生前所嘱，大兴土木而造。这一批建筑，皆为规模宏大、造型优美的中西合璧之作，而华南理工大学建筑领域的杰出人才辈出，所以此片土壤，亦是建筑的试验田和梦工场，新旧建筑都很有看头。

1	3
2	4

1-2.中山大学石牌旧址建筑群的红砖楼掩映在葳蕤草木中。
3.中山大学石牌旧址建筑群正门牌坊。
4.全国第一个砖拱薄壳建筑——红满堂。

中国人自己规划建设的山水校园

中山大学石牌旧址建筑群

斗壁飞檐,红墙绿瓦,掩映于花间与湖畔,林荫校道穿连其中,不愧为近代中国人自己规划建设的中国式山水校园。

年代: 近代
地址: 天河区五山华南农业大学、华南理工大学

红墙绿意,琉璃瓦当,错落有致,洋紫荆、樱花、茶花、羊蹄甲、黄花风铃木等次第盛放,五山的华工华农双校园,引长枪短炮四季不绝,如梭人潮内又有多少客知道,目之所及的旧建筑群,多半是中大石牌旧址的校舍。

1924年初,孙中山先生以陆海军大元帅之名,下令创办一武一文两所学校,武是黄埔军校,文是由当时广东三所高校组建而成的国立广东大学,这是中国华南地区第一所由中国人自己创办的多学科大学,鲁迅、郭沫若、丁颖等学者先后到校任教。按先生遗嘱,1932—1937年间在广州东郊石牌地区(现华南理工大学及华南农业大学所在地)规划建设了中大新校园,成为中国近代一所由中国人规划建设的、中国式的高校山水校园,在中国近代教育史和建筑史上都具有特殊价值。

最早参与建校的设计师大都有留洋背景,因此校园初建时就具备良好的规划,从荒芜之山地到"藏修息游"的校舍,营建计划分三期完成。

第一期工程由杨锡宗主持,他与当时中大校长邹鲁共同完成总体规划,其一大特色是"钟"形的道路设计,此创意源自地形,又因地形而夭折,但在旧址上处处可见的"钟"字图腾,透露出规划时的用意;第二期在林克明主持下,校园主体建筑成形,建筑物以仿明宫殿式样的传统风格为主,细部构造上借鉴西方要素,牌坊、亭子、日晷等构造物尽显中国传统特征,校园景观风貌则是几何式与自由式并存;第三期由余清江主持。

邹鲁在1928年前后考察了29个国家及其高等教育和校园建设,对国立中大的办学及石牌校园的建设起了积极重要的作用。今天,斗壁飞檐之教学楼,红墙绿瓦之宿舍楼,掩映在绿树中或湖岸边,林荫校道穿连其中,的确"嘉惠学子不浅"。

Building Complex of Former National Sun Yat-sen University (Shipai Campus)

Year Built: In modern times
Add: South China Agricultural University, Wushan, Tianhe District

The campus lies by the lake amid blooming flowers, featuring overhanging eaves, red walls, and shaded boulevards. It is a modern Chinese-style landscaped campus, designed by Chinese designers.

1	2		4	5	6	7
		3				

1—2.红砖绿琉璃、斗壁飞檐,融贯东西的建筑语言,虽近百岁,仍见端庄秀美。
3.入春时5号楼楼前樱花、茶花盛放,也是爱花的广州人最爱的赏花地。
4—7.细节之美。

第三章│余芬
Chapter 3:Helping the posterity

华南理工大学建筑群

年代： 近代
地址： 天河区五山华南理工大学

知识的殿堂，理性的韵律

沿南北中轴从华农游到华工，新旧交错之间，昔日校园中心区的规划依然清晰可见，原中大的老建筑与华南工学院时期建造的现代岭南建筑，共同见证着近一个世纪的时代巨变。

在巨变的洪流中，岭南建筑学派逐渐形成。1945年夏昌世出任华南理工大学前身——中山大学建筑工程学院主任，在他的努力下，华工陆续出现了一系列紧扣时代、经济实用、"智慧与美貌并重"的建筑。而"夏氏遮阳"等建筑学上的新思潮新技术在华工的竞相盛放，如春日灿烂的洋紫荆，散发着南国特有的迷人光彩。

Building Complexes of Former National South China University of Technology

Year Built: In modern times
Add: South China University of Technology, Wushan, Tianhe District

The campus sees Bauhinia in full bloom. Its buildings of modern Lingnan style built back in the period of South China College of Engineering, together with the old buildings of the former National Sun Yat-sen University, have witnessed almost a century of vicissitude.

113

1	3
2	4

1. 建筑系办公楼常被人们亲切地称作"建筑红楼",华工人饱含深情地谓此处美景作——"红楼映翠",仿中国古典建筑风格,绿琉璃瓦黄屋脊红墙钢窗……细节精巧、气势恢宏。
2. 建于1936年的体育馆为民族风格的宫殿式建筑,中式的牌坊式冲天柱、牌楼式屋顶间隔排列,西式拱门、中式勾栏,兼具东方的华美与西方的流丽,端庄中又见浑厚。
3. 著名建筑师杨锡宗设计的9号楼,现在是华南理工大学电力学院,入口门廊抱厦,整栋建以合院式布局围合,形体流畅而周正。
4. 气势恢宏的文学院,红墙绿瓦中西合璧宫殿式建筑,门廊的4根西式纺锤形巨柱,颇有气吞万象的气势。红砖墙、花岗石石脚、孔雀脊饰,兼东西方之殿堂庙宇肃穆巍峨之气象,这也许是全国最具豪迈峻朗之气的文学院。

In this palace of knowledge, Chinese and Western buildings in varied forms and styles blend harmoniously, creating a rhythm of rationality.

In South China of the 1950's, Xia Changshi redefined the characteristics of Lingnan architecture. Along with the implementation of the Xia-style Shading approach, a series of cost-effective, smart yet aesthetically pleasing buildings were gradually constructed on the campus of South China University of Technology.

1	2	3		7	8	9
4	5	6			10	

1.建筑红楼的主楼有14根巨大的圆形红色檐柱，配以上方的彩绘梁衬，富丽堂皇。
2.1号楼位于校园中轴线上，正对孙中山雕像。
3.体育馆的玻璃窗户饰有高浮雕花纹，简约而耐看。
4.脊兽常有惊喜，羊、孔雀……题材大胆跳脱。
5.工学院机械电气工程教室构建虽受西方建筑的影响，屋顶歇山顶、正脊、兽们和大门两旁的须弥座、栏杆的雕饰、抱鼓石及檐下的双重桁饰等，都吸收了中国古建筑的精华。
6.遮阳又挡雨兼具装饰性，华工人，无所不试。
7—9.华南理工大学建校之始，在建筑领域人才济济，常有让人耳目一新的建筑语言与手段亮相世人之前。
10.正午法学院前，1936年启用的日晷，投影分秒不差。

夏昌世1952年接手设计图书馆，以平屋顶取代了宫殿式坡屋面，将主入口设于南面，通过宽阔的台阶导人流至二层，浮现学术殿堂之感，华工图书馆是这批建筑中的杰作，1936年杨锡宗的原设计为仿中国古典宫殿式建筑，因抗战而停工，夏昌世接手已是1952年，他将其改造成一个经济实用、适应南方气候且外观新颖的图书馆，一度成为国内图书馆的典范。以平屋顶取代坡屋面以节省经费；中部增设两个内天井，以宽敞的走廊纵横贯通，带进穿堂风将内部热气从天井压出，为大进深空间"天然降温"。

同样独具一格的还有华南工学院3号楼（自动化学院），"U"形平面，外廊式与内廊式组合结构，综合式遮阳方法，屋顶拱形隔热设计等，皆是"经济实用"的建筑实践。后来"夏氏遮阳"在华工1号楼的运用更是精彩，不仅满足基本遮阳，立面也形成了理性的韵律感，是"美貌与智慧"并重的佳作。

在中华人民共和国成立之初，百废待兴，正是夏昌世和同道一起，在遮阳、隔热和通风的解决方案中同时使用了新技术和乡土技术，挣脱了形制与样式的束缚，将中国传统的园林精神引入岭南现代建筑，让空间自由流动，从此岭南建筑被赋予了新的性格，理性而富有意蕴。

在校园行走，但见中国传统园林精神与岭南现代建筑风格碰撞又相融，开放式的山水高校，继承着前人的雄心壮志，回荡近百年的一曲满园春，仍在耳畔哼唱。

一条中轴线，一页东西史
Early Building Cluster of Kangle Garden
康乐园早期建筑群

Early Building Cluster of Kangle Garden

Year Built: Early 20ᵗʰ Century
Add: Kangle Village, Xingang Lu, Haizhu District (The present-day Sun Yat-Sen University)

Kangle Garden of Sun Yat-Sen University, with its carpets of green grass and lush banyan trees, is ranked as one of the three most beautiful Chinese campuses, together with Weiming Lake of Peking University and Luojia Mountain of Wuhan University. In the Kangle Garden, the architecture is designed to reflect the university's tenet, while the vegetation enhances the environment's attractiveness and stimulates student activity specimen of the architectural ideological trend of that specific period.

一条中轴线，一页东西史

康乐园早期建筑群

康乐园的精气神，以建筑立精神，以植被舒气韵，以琅琅书声传神秀。其早期建筑群更是凝固了一个时期的思潮动向，是建筑史上的活标本。

年代 20世纪初
地址 海珠区新港路康乐村（现中山大学校址）

　　珠江南岸边中山大学南校区，又称康乐园，与北大未名湖、武大珞珈山，并称中国大学三大最美校园。

　　当年宋文帝曾将晋室贵族、中国山水诗创始人康乐公谢灵运流放广州马岗顶，此地因此得名"康乐村""康乐园"。

　　20世纪初，格致书院看中了康乐园的幽静，定址迁入，建立岭南大学，及后，岭南大学再演变成中山大学。

　　岭南大学从图纸时期开始就弥漫着西方近代大学的布局气息。一条中轴线贯穿南北，各教学建筑一律南北朝向，轴对称排布。不以一进一进的院子为单位，因而便没有形制上的主次，每栋建筑构成一个"独立王国"。以捐款人之名为建筑命名的风气也是相当洋派。

　　建筑物之间的相对独立，获得捐赠的时间又各不相同，不同风格和理念的建筑师用他们的构想，使康乐村的早期建筑群呈现出了"和而不同"的独特风貌。

　　中轴线上的怀士堂，又称"小礼堂"，孙中山先生和宋庆龄多次到岭南大学视察，两人曾于怀士堂前合影留念，中山先生在怀士堂长篇演讲勉励青年学生"立志，是要做大事，不可要做大官"，余音绕梁。

1	2	3	8
4	5	6	
7			

1—6. 红砖可算康乐园内的建筑小品，三顺一丁、一顺一丁、梅花丁、全顺全丁，各种组砌方法谱成一曲曲红砖小调。
7. 格兰堂。
8. 时间的余荫，岁月的余芬。

康乐园东南区一号,语言学家、历史学家陈寅恪生命的最后20年就住在这里。患目疾的陈教授在这栋两层小红楼里给学生讲课,这里是人们心中的"金明馆""寒柳堂",瞻仰者经年不歇。

能与之媲美的,是小礼堂东面的黑石屋,这里曾是岭南大学首位华人校长钟荣光博士的寓所,现是学校贵宾招待所,当年德国总理施罗德到中山大学参观访问便下榻于此。

黑石屋、马丁堂作为"中西合璧"早期阶段的代表,建筑整体造型偏西式,仅带少量中式建筑符号。比如屋顶均为西式直坡面,但会采用中式悬山顶或歇山顶。中西元素,直来直往,短兵相接。

在过渡期阶段诞生的十友堂、爪哇堂等,则更加注重建筑的实用性,根据不同功能将多个坡屋顶进行组合,高低错落,参差别致。中西交融逐渐从表皮渗透进肌理。

20世纪20年代后期到30年代属于成熟期。随着岭南大学开始由国人自办,这一时期的建筑不再以西式建筑特征为主导,而是表现出更为明显的中国传统样式,比如陆佑堂和哲生堂都采用了仿清代宫廷式建筑形制,有了黄色的外立面、红色的廊柱、雕梁画栋和站着神兽的飞翘屋檐。

从岭南大学到中山大学,康乐园的规模渐大,但原有的中轴清晰明朗,从孙中山铜像、惺亭、乙丑进士牌坊,数十亩绿茵地一路延伸到岭南堂,校道两旁榕荫蔽日、鲜花长放、建筑端丽、学子如织,依旧是当年育做大事之人的大气象之地。

1. 爪哇堂。
2. 八角亭。
3. 陆佑堂。
4. 哲生堂。
5. 老樟织出满园浓荫。
6—7. 陈寅恪故居。此楼楼高两层,为美国麻金墨夫人为纪念其丈夫于1911年所捐建,故名"麻金墨屋",《论再生缘》《柳如是别传》等著作就是在此写就。

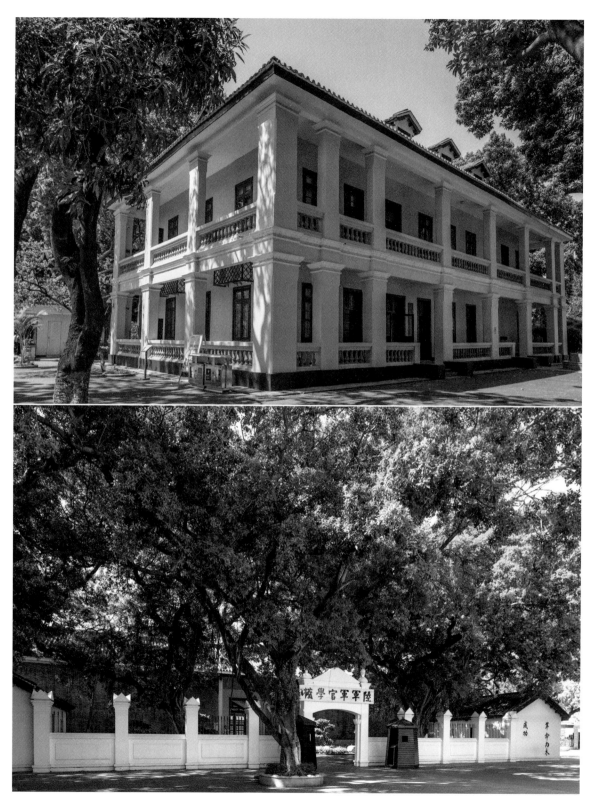

Former Site of Huangpu Military Academy

Year Built: 1924
Add: Inside Changzhou Island, Huangpu District

The academy was not only the cradle of Chinese generals but also one of the four renowned military schools in world history. The remaining school gate, Sun Yat-Sen's former residence and monument, as well as theater-style club building are still considered as a pilgrimage site by numerous overseas Chinese.

这个名字，曾又叫作赤诚、勇气与肝胆相照

黄埔军校旧址

这是中国将帅的摇篮，是近代最为著名的军事学校。多少心怀救国志的少年从挂着"升官发财请往他处，贪生畏死勿入斯门"对联的牌坊昂首步入，开始书写自己的人生传奇。

年代 1924
地址 黄埔区长洲岛内

| 1 | 3 | 4 |
| 2 | | |

1—4.对海内外华人而言，黄埔军校地位崇高。这里培育的英才，曾经大大改变了一个国家的格局。

传奇学堂黄埔军校之所以在黄埔长洲岛落地，首要还是地利。

长洲岛扼珠江之咽——从外海进入狮子洋，再西溯由珠江入城，长洲岛是必经关隘。清代长洲炮台群，长洲岛位居核心。

自清代开始，岛上陆续兴办了旨在培养国防人才的博学馆、水师学堂、陆军小学、海军学校，校舍、校区仍可沿用，教育基础良好。长洲岛与对望的黄埔古埗，一后一先成为清代广州口岸一口通商时代的外贸船只的挂单口，长洲岛因外航船只常年停靠，船舶制造和修理业渐成规模，有船坚炮利的技术支撑。

小岛自成一体，不受各部势力干扰，学员有安心学业的小环境；岛上民众眼界早开，易于接受新事物，又有军民融洽的小氛围，种种良因终造佳话，这所中国将帅的摇篮，于1924年6月在黄埔长洲岛正式开学。

八卦山上的孙中山纪念碑，是校区主体区域最高的构筑物，从珠江航行远远便可望见。碑上铜像朝北远望，暗含先生北定中原、统一祖国的未竟志愿。

黄埔军校的主体建筑背靠山岗、面朝珠江，按岭南民居祠堂式的中轴对称式布列，共有三路四进，每进各自围出4个天井，既保证良好的采光，又各自留出了活动空间。

因广州炎热多雨的气候，黄埔军校的建筑主体以连廊相连贯穿，既遮阳又挡雨；各房皆为南北通，通风和采光皆良好；墙身砌青砖以防潮防蚀，屋顶倾角加大更利倾泻雨水。整体建筑风格朴实克制的黄埔军校，又于细微处做到经世致用，中式庭院布局、体量恢宏的黄埔军校，自有一种沉稳的威严庄重。

Guangzhou Luxun Memorial Hall

Year Built: 1904
Add: 215, Wenming Lu, Yuexiu District

The bell tower witnessed the first Kuomingtang-Communist Party cooperation and once served as Lu Xun's residence when he was teaching in Sun Yat-Sen University. There are many stories yet to be told about it.

鲁迅住过的钟楼，曾奏响中国近代史的光辉乐章

广州鲁迅纪念馆

第一次国共合作从这里启程，这里还留有广东贡院号舍基址，鲁迅在中大任教时，也在此居住。

年代 1904
地址 越秀区文明路 215 号

1	2	
	3	4

1. 中国国民党第一次全国代表大会旧址，是在广东贡院旧址重建，也是中山大学最早的校址。
2. 首层的礼堂，保持着孙中山先生主持中国国民党第一次全国代表大会时的原状。
3. 小礼堂的长凳上，曾经端坐过许多风云人物。
4. 鲁迅先生在中大任职时曾入住此处。

这座仿罗马古典式砖木结构建筑群当中，鹅黄色钟楼尤其显眼。门口挂着两块牌，右侧是"中国国民党一大会议旧址"，左侧是"广州鲁迅纪念馆"。

入内见一300多平方米小礼堂，此处原为清末广东贡院所在，康有为、梁启超都曾在此参加乡试。1912年，这里被改为国立广东高等师范学校；1924年，孙中山在这里主持召开了中国国民党第一次全国代表大会，并将此改名为广东大学；后为纪念孙中山又更名为中山大学，这钟楼便是中大最初的办公楼，亦是中大校徽的标志性图案。

站在礼堂内，耳边广播传来孙中山先生讲话原声，眼前是当年国民党一大会场原样摆设。主席台布置庄严，戎装的孙中山画像被挂在正中墙面上，而标有繁体字"三十九号毛泽东"的木凳就在左起第三排。

时钟往后拨一点，1927年鲁迅在中大任教时便住在钟楼二楼，如今钟楼内鲁迅纪念厅重现了这位文学巨匠从出生到日本留学再到广州参与革命的历史点滴。走出钟楼，门前是芳草茵茵的革命广场，有写着鲁迅语录的石椅。

留有如此多重要的人与事的旧痕，这是一栋被时光祝福过的房子。

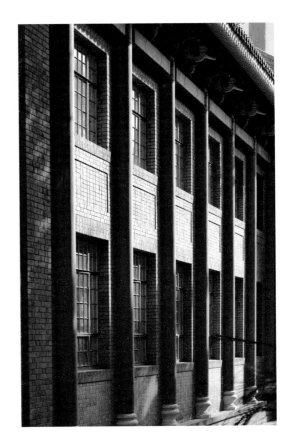

无问东西之岭南建筑派，
林克明设计的现代藏书楼

广州市立中山图书馆旧址

这座广州近代第一座专设的公共图书馆，红墙绿瓦，中西合璧，隔绝了尘嚣，留存了文脉。

年代 1933
地址 越秀区文德北路 81 号

广州市立中山图书馆由华侨为纪念孙中山先生捐款兴建，于1933年10月建成，是广州近代第一座专设的公共图书馆，现在是广东省立中山图书馆少儿部的所在。

它是岭南建筑大师林克明留法归国后的第一个作品，建筑外形为传统宫殿式样，内部则用材和结构都采用西方先进技术。这种"中学为体、西学为用"、保持民族固有形式的建筑风格，是广州近代建筑中的一大潮流。

广州自古的文脉也正在此地，曾筑有最高学府——广府学宫，番山遗址更位列明清羊城八景，完好保存至今的翰墨池相传为千年古迹。步上番山亭，如登书山路，昔日青年建筑师探索实践建筑新风的身影，鼓舞着后继之人。

Site of Sun Yat-Sen Library of Guangzhou City

Year Built: 1933
Add: 81, Wende Beilu, Yuexiu District

Being the first dedicated public library in modern Guangzhou, the Library was designed by Lin Keming from the Lingnan architecture school. It features red bricks and green tiles, typical of traditional Chinese architecture, while also incorporating elements of Western architecture, thus representing a fusion of the two architectural styles.

1	4	5
2	3	

1—2.虽是框架结构、钢筋水泥，但东方的神韵被完好保存。
3.华侨为纪念孙中山先生兴建的这处公共图书馆，至今仍作为少儿图书馆正常使用。
4.番山之上的番山亭，广州的两座小山丘，番山和禺山，也是旧称番禺的来由。
5.80载春秋，番山之下，华夏巍峨。

Mingxin Academy

Year Built: 1912
Add: 5, Mingxin Lu, Liwan District

With its colonial-style external corridor, plain red wall and western-style clay sculpture, the Academy is the first barrier-free school for the blind in Guangzhou. The sense of order is evident in both its function and architectural design.

从白鹅潭边明心之地，看建筑的秩序

明心书院 /

殖民地外廊式风格，清水红墙配西式灰塑，这里是广州最早采用无障碍设计的盲人学校，从功能到建筑设计，处处见"秩序"。

年代 1912
地址 荔湾区明心路 5 号

白鹅潭畔，花卉芬芳，芳村一带在清末时吸引了众多西方传教士前来办学，明心书院，便是由广州博济医院美籍华人医师冯西所创办。清水红墙，砖木混合结构，殖民地外廊式风格，造型简洁朴实。首层的拱券廊、楼顶的女儿墙、精美的西式图案灰塑，历经百年洗礼而风韵未减。

明心书院是广州第一所盲人学校，也是广州最早采用无障碍设计的盲人学校，其人性化在功能使用与建筑设计方面均有体现。充分考虑到盲人的使用习惯，全部采取了无门槛式的无障碍处理，这在当时的建筑中极罕见；二、三层房间则全铺木地板，使室内平面与外廊地板平齐，以保证盲人行走方便。

书院旧址主教学楼坐西朝东，在东、南两面设外廊，其中东立面各层均为连续拱券构成的券廊，南立面则首层为廊柱，二、三层为券廊，井井有条；西、北两面为红砖实墙，开有竖窗，窗上设有弯拱窗楣，精巧雅致。所有外廊均设有水泥通花栏杆，铺就木地板，毫不含糊。这些设计细节，使得整体建筑的造型线条简洁明快，而秩序感也由此得以强化。

从最初简陋的盲人女塾，到鼎盛时期的100多名盲童在学，明心书院赋予这些与众不同的学子们谋生的本领。目虽盲而心明白，建筑的秩序，这里的使用者有着最高话语权。

1	2		4	5
	3			

1—5.明心书院旧址的6栋朴素旧筑，体量较大，格局完整，虽不算华美，但它们的背后，有一个温暖的故事。

第四章：商都
Chapter 4: City of Commerce

作为货如轮运、往来频仍的千年商都，广州很早就迎接了海洋文明的滋养与考验，商贸活动中催生的各种建筑配套，记录着商都广州永不落幕的繁华。你可以想象，当年越洋的冒险家看到珠江三塔时，心里会对广州泛起何等温暖的涟漪。

Influenced by the marine culture in the past millenniums, Guangzhou has witnessed the rise of various buildings that support its business activities. These buildings vividly documented the everlasting prosperity of the city as a millennial business hub.

第四章 | 商都
Chapter 4: City of commerce

| 1 | 4 | 5 |
| 2 | 3 | |

1. 飞檐衬映一轮新月，落霞时分，尤为静穆祥和。
2—3. 从兰圃往西，经过这道四柱三门，便又是一片宁静之所。
4. 作为海上丝绸之路的重要遗迹，此处地位尊崇。
5. 来礼拜先贤墓园的各国游客络绎不绝。

Tomb of Saad ibn Abi Waqqas

Year Built: From Tang to Qing Dynasties
Add: 901, Jiefang Beilu, Yuexiu District

With a worship hall for 3,000 worshippers, the Tomb of Saad ibn Abi Waqqas is now the largest venue for religious activities of Islam in Guangzhou.

逝者给生者的最后一课，
贤与非贤在永恒里相伴而眠

清真先贤古墓

伊斯兰教圣人之墓，礼拜堂可供 3000 人共同礼拜，是广州现今最大的伊斯兰宗教活动场所。

年代 唐—清
地址 越秀区解放北路 901 号

　　清真先贤古墓的第一道石牌坊离马路很近，四柱三门，朱砂大字——"先贤古墓道"。两侧的耳门上是阿文圣训，"死亡足以发人深省"，"现世足以分晓逝迁"。两种文化手拉手把喧嚣的尘世挡在身外。提醒人们，接下来的课程，将由逝者教授。

137

Immersed in the fragrance of osmanthus, the pathway winds its way to the Saad ibn Abi Waqqas Mosque, Mausoleum of the uncle of Prophet Muhammad.

然后一路上坡，非颔首不足以酝酿敬重之心。路终点是礼拜堂，飞檐衬新月，绿瓦遮空殿，又是两种文化的相生共荣。随着外陵版图的次第展开，伊斯兰教传教士艾卜·宛葛素的故事渐渐清晰。相传这位先贤辈分极高，乃穆罕默德之舅。唐初来广州传教，后多方游历，被尊为贤能者，晚年回到广州并长眠于此。当代许多伊斯兰国家的首脑抵穗无不前来拜谒。

正因身份贵重，宛葛素的墓园才把"得见真容"的前路铺设得层层递进，行墓道、拜礼堂、过二道牌坊，还要再走过一条墓道，穿过41位贤者的坟墓才能真正到达宛葛素墓室——保敦和堂。此时仪式感已被推至顶峰，再看状如钟罩的墓室才更觉庄严。穹隆顶在视觉上模拟了苍穹。宛葛素石棺高高隆起，上覆五彩锦缎。诵经时，层层声波叠加，微音也成大吕。加之室内异香经年不散，色身香味触法、眼耳鼻舌身意，凡心再炽也能静下来。

先贤高山仰止，环绕而葬的41位贤者的故事则是传奇。相传有个屠夫，不信穆斯林如传说中一样虔诚坚定（穆斯林在做礼拜的时候不能干其他事，纵使生命危矣亦不可动），于是在宛葛素的40随员做礼拜时，逐个将他们砍杀，竟无一反抗（一说强盗抢劫）。屠夫幡然悔悟，当即自刎。后人遂把他和他的"受害人们"葬在一起，不贤但能改，亦和殉道者同等待遇。

用当代世俗观点来看，这真不是一个公平的故事。但强化仇恨就公平了吗？宗教以其博大的善，用死的故事探求更生的解决方案。就像院内的一处墓碑碑文："当你处世彷徨时，便求教于墓中人吧！墓中有天堂的乐园，也有地狱的火坑。"

1. 穿过这道门，便是伊斯兰教圣贤穆罕默德之舅——宛葛素之墓。
2. 墓道遍植桂花，愈深幽愈见芳香。
3—5. 宛葛素之唐代古墓之后，榕已成荫。

商都心脏，
千年不变中轴线

千年古道遗址

经历多朝变幻，千年古道如今安静地躺于玻璃罩下，在时间的河流里，商业和文化无分你我。

年代 唐—清
地址 越秀区北京路

　　秦汉时，广州城内有番山和禺山两座山岗，故称番禺城。唐末开拓城区时，凿平山岗而筑出千年驿道，便是现在的北京路。

　　2002年夏，广州市政府对北京步行街大整修，挖开路面后，古迹浮现，出土了大量的砂岩石条和古城墙墙砖。经考证，古道始于唐代，出土的砂岩石条是自5个朝代11层的路基路面。

　　旧时北京路，北有四方炮台高岗、官府，南有双门底、码头，是由南进入广州城的官方通道，也是广州在唐宋元明清时期的城市中轴线，千年不变。

Site of Millennial Ancient Road

Year Built: From Tang to Qing Dynasties
Add: Beijing Lu, Yuexiu District

Despite ups and downs of the changing dynasties, the millennial ancient road now lies quietly beneath the glass cover. It remains the forever central axis of the city and the ever-beating heart of the business hub.

3	
1	
2	

1. 隔着玻璃观看宋代路面。
2. 宋代门洞路面。
3. 宋代路面地砖。

第四章 | 商都
Chapter 4: City of commerce

| | 2 | 3 |
|1| 4 | |

1. 琶洲塔为穿壁绕平座式结构，要穿过塔壁出去平座，绕过平座才能继续登上一层楼；红砂岩垒筑基座，基面以灰色斑岩铺砌，基侧分别刻有八卦图形，显示此塔为风水塔而非佛塔。
2. 莲花塔的塔刹由覆盆、宝珠、仰莲和铜葫芦组成，阳光照耀下流光溢彩，数十里之遥举目可见，海外来舶望塔而知广州至。
3—4. 赤岗塔红白相间的俊秀塔姿，在绿树红花的簇拥中，别有一番岭南风味。

风水引航三宝塔，
看尽海上丝路繁华

琶洲塔 | 莲花塔 | 赤岗塔

这三座风水宝塔，锁二江、束海口，雄踞珠江400年，是当年船舶从水路进入广州城的航标。

年代 明代
地址 琶洲塔：海珠区琶洲新港东路
　　　莲花塔：番禺区莲花山
　　　赤岗塔：海珠区新港中路艺苑路

琶洲塔、莲花塔、赤岗塔兴建于明代，昔日广州有诗云：白云越秀翠城邑，三塔三关锁珠江。明清时期，来到珠江口的远洋商船，看到这三塔便知前方到广州城。

古人深信，这三座雄伟的"风水宝塔"，巍然耸立在珠江边，可关锁水口，若说羊城是一艘乘风破浪的大帆船，那么，雄踞珠江的三塔，则象征着帆船上的三支桅杆，庇佑着全城风调雨顺，事事顺利。

400多岁的琶洲塔，八角形楼阁式、青砖砌筑，屹立于江中小岛形如琶琶相连的山岗上，像极中流砥柱，于是有了清代羊城八景之"琶洲砥柱"。

莲花塔也是明代砖塔，粉墙红柱、绿琉璃瓦、八角攒尖顶，塔内层层铺设木楼板，各级以砖牙挑出瓦檐。因坐落在莲花山上，是从水路进广州见到的第一座塔。

赤岗塔与琶洲塔相似，塔下岗土岩石皆呈红色，故名赤岗，塔基八角均镶有番方人形象的托塔力士石雕佳作，是研究明代石雕与广州海外贸易的重要实物资料。

阅尽沧桑数百年，三塔虽卸下指示航道的重任，但琶洲建了新的国际会展中心，赤岗建成领馆区，莲花山上莲花塔成了国际旅游景点，三塔雄风依旧，继续招商待客。

Pazhou Pagoda / Lotus Pagoda / Chigang Pagoda

Year Built: Ming Dynasty
Pazhou Pagoda: Xingang Donglu, Haizhu District
Lotus Pagoda: Lotus Mountain, Panyu District
Chigang Pagoda: Yiyuan Lu, Xingang Zhonglu, Haizhu District

Towering over the Pearl River for 400 years, the three Fengshui pagodas used to function as beacons for passing ships, bearing witness to the prosperity of the Maritime Silk Road in the past millennia.

海上丝路的"丝"有多辉煌，
线索在这里
Jinlun Guild Hall
锦纶会馆

海上丝路的"丝"有多辉煌，
线索在这里

锦纶会馆

锦纶会馆是广州现存的、结构最为完整的清代行业会馆，存于馆内的22块碑记，记录了广州的海上丝绸之路上丝织行业的辉煌历程。正因锦纶会馆的重要意义，广州人将它整栋平移到现址，开创世界首例砖木结构古建平移的先例。

年代 清初
地址 荔湾区康王南路289号

1	3	4
2	5	6

1. 锦纶会馆的平移，开创了世界先例。
2—4. 因年久失修，锦纶会馆的灰塑、陶塑多有损坏，以别处旧件或按旧形制修旧如旧补充。
5. 第三进的花格采光天窗，是旧时明瓦的做法，在花格中嵌进磨至透明的蚝壳薄片，以达到较好的采光效果。
6. 当年加入锦纶会馆的纺织业商家，多达400家之众。

Jinlun Guild Hall

Year Built: Qing Dynasty
Add: 289, Kangwang Nanlu, Liwan District

Built in the Qing Dynasty, Jinlun Guild Hall has the best-preserved building structure among its counterparts in Guangzhou. It housed 22 inscription tablets that record the glorious history of the silk weaving industry in Guangzhou as part of the Maritime Silk Road. With a view to the profound significance of the Hall, it was relocated as a whole to its current location, marking the first time in the world an ancient building of brick-and-wood structure has been moved in this way.

2001年广州城西扩路，为了保留一栋珍贵的建筑，广州人开创了全世界第一例砖木结构古建平移的先例，整栋建筑向北平移后再抬高、再向西平移，又再花了三年多时间修旧如旧，终将这栋建筑以昔日风貌呈现人前。

这栋近300岁的建筑，便是广州市唯一得以完整保存的清代行业会馆——锦纶会馆。

锦纶会馆于雍正元年（1723年）初创，此时虽离广州"一口通商"时代尚有30余年，但全国各地的口岸税收，还是广州一家独大。凡广州口岸出口的商品，丝制品比重很大。因外贸需求庞大，广州、佛山的织工，有数万之众，单是加入锦纶会馆的织业店铺就有400家。

一口通商后因江南丝织品只能从内陆内江辗转运抵广州，货期长、成本高，所以一度珠三角遍布桑基鱼塘，自产蚕丝，自缫自织以使货美价优的丝织品产业链利润丰厚、货如轮转。

产业链上产、供、销诸事千头万绪，行业组织作用显得尤为重要，锦纶会馆亦可谓绵延至清代的海上丝绸之路，极为重要的遗痕。

今天三路三进的锦纶会馆，完好地保留了22块碑记，记录当年数百名织业东主选值事、议行事、拜祖师、赏戏文、会馆重修与扩建等大小诸事，年代最近的一块碑是1924年所立的《重建会馆碑记》，碑记为廖仲恺所题，记录了孙中山先生"永远不得别立名目"的指示。当年国民政府财政捉襟见肘，一度大面积抛售充公公财，锦纶会馆亦在此列，所幸得孙先生的指示，才为后人保下了这一处意义重大的会馆。

从前，
老广的一天从这里开始

粤海关旧址

游船观珠江两岸景致，这座欧式古典钟楼尤其显眼，罕见的英制全机械传动式立钟，是一代广州人的钟楼记忆。

年代 1914
地址 荔湾区沿江西路 29 号

粤海关与闽海关、江海关、浙海关合称清初四大海关。

粤海关旧址大楼于1914年奠基、1916年建成，由英国建筑师大卫·迪克设计，其欧洲新古典主义建筑风格浓厚，是近现代重要史迹及代表性建筑。

因建筑顶部立有穹顶钟楼，故俗称大钟楼。钟楼里有目前全国罕见的、保存完好的英制全机械传动式立钟，往时广州人便以此钟作为标准时间参考，这也是很多广州老街坊的童年记忆。

新中国成立后至 2006年，此处一直作为广州海关的办公地。昔日钟声虽不再，钟楼建筑庄严典雅，也是老广招待来客游珠江时拿得出手的名片。

Site of Former Guangdong Customs

Year Built: 1914
Add: 29, Yanjiang Xilu, Liwan District

In old days, this classic European-style clock tower reminded the Cantonese of the beginning of a new day. Now, it is the most eye-catching icon for tourists of the Pearl River cruise. This British-made clock of full mechanical transmission stands as the memory of a generation of Guangzhou residents.

第五章：人居
Chapter 5: Houses

广州人的人居智慧与审美，向来是在南来北往、东西融会的海纳百川之后，再转化成自己的完善体系，广州人的人居，如南国四时盛放的花朵，绮丽华美，一派富丽安康的气象。

As a crossroad where the southern and northern Chinese culture converges and the East meets the West, Guangzhou has developed its own wisdom and aesthetics of living. The houses in the city are gorgeous, magnificent and peaceful, just like the flowers that bloom throughout the year.

这种房子，
曾是财富与地位的代名词

西关大屋建筑群/

青云巷、趟栊门、满洲窗、硬木桌椅、西洋古玩……昔日西关富商的生活细节重现。

年代 清代

地址 荔湾区龙津西路逢源北街 84 号（荔湾博物馆）

　　西关文化，是广州文化的缩影；而西关大屋，又是西关建筑文化的精髓。

　　西关大屋是建于清中晚期的高档民居，其平面布局是基于传统三间两廊的基础上演变而来，平面沿纵深方向排列若干进房间及天井，左右分别布置侧室、楼梯间及厨房。这种格局，功能分区明确，内外界线严明，显出主人的身份。在今荔湾博物馆一窥西关大屋的格局与布置，当时居民的生活形态便一一重现。

Xiguan Mansions

Year Built: Qing Dynasty
Add: 84, Fengyuan Beijie, Longjin Xilu, Liwan District (Liwan Museum)

These mansions used to be the synonym of wealth and social status. Grey-green brick lane, Tanglong door and Manchuria window, and rosewood furniture used to be symbols of rich businessmen's daily lives.

| 1 | 2 |
| | 3 |

1. 西关大屋楼层较高，以利拔风，而满屋是做工华美的硬木家具，一室富丽景象。
2. 房间里以嵌彩色玻璃的窗格或明瓦天窗来采光。
3. 院里常有小花园，一利采光，二可供莳花弄草，陶冶情操。

第五章 | 人居
Chapter 5: Houses

一条短巷，
百段传奇

耀华大街

耀华大街——这条狭窄的麻石巷，是广州现存最具代表性的西关民居聚落，西方的装饰主义与建筑技术，毫无违和感地与东方的人居智慧相融合。

年代 清代
地址 荔湾区耀华大街

广州西关小巷纵横，耀华大街是广州最具代表性的传统民居聚集区，是石板铺筑的麻石巷，巷子两侧主要为西关传统民居建筑，外地游客想看传说中的西关大屋"三件头"——脚门、趟栊门及板门，都会先跑来这里。

大街东西走向，宽不过丈，短短100多米，30幢有门牌的房屋有多处被挂牌"历史建筑"，间或有碑刻画壁点缀，花木扶疏之间，几乎每座老房子都有段"古"。

细看每家阳台栏板和檐口的纹饰都不甚相同，雕花的栏杆、考究的窗子，各有志趣。绝大多数建筑都保留了石门洞、石墙脚，随便走进一家，门前都铺设有石板台阶，大户的房子更有近1米高的门槛，气派犹在。

当年老宅的主人，多为粤剧艺人、华侨、医生、商人、政要，广州花纱布业巨头蔡氏家族，著名粤剧名伶白玉堂、郑绮文等都曾是"街坊"。而如今在此生活的新广州人，与老广们和睦相处，在这宁静一隅，人间烟火的节奏丝毫没被过客扰乱。

1.耀华大街东起文昌北路，西至耀华西街的西关大屋街，有房屋30多幢，皆为民居，是广州目前西关大屋最集中的地方，是重点保护的街区。
2.矮脚门、趟栊门、板门，是西关大屋"三件头"。
3.麻石板铺成的路。

Yaohua Dajie

Year Built: Qing Dynasty
Add: Yaohua Dajie, Liwan District

This narrow stone-paved lane is the most representative element in Xiguan residential settlement that can still be found in Guangzhou. Here one may see how the Art Deco and architectural techniques from the West are seamlessly combined with the wisdom of living from the East.

155

时光的廊，
广州最完整最长的骑楼街

恩宁路骑楼

一条街串起了多处文物古迹，楼上住家，楼下经商，因地制宜，是广州人的智慧。

年代 20 世纪早期
地址 荔湾区恩宁路

恩宁路是一条有浓厚西关特色的道路，是西关骑楼建筑的精髓。骑楼建筑由近代时期，从西方古典建筑与岭南传统建筑相结合演变而来，其上居下商的空间方式，满足了商业与居住相结合的需要，底部架空的空间处理方式很好地适应了本地的气候条件。

恩宁路之所以取名"恩宁"二字，据考是过去恩宁路的一头有一条恩洲村，另一头是一条宁溪村。修建这条路时，必须穿过这两条村，后经双方协议，最终就命名为"恩宁"。

恩宁路与龙津西路、第十甫、上下九步行街骑楼连接，成为全市最长、最完整的骑楼街，分布着八和会馆、李小龙祖居、詹天佑纪念馆、泰华楼、金声电影院等十几处文物古迹。沿恩宁路漫溯，满耳听不完岁月的吟唱，满目读不完时光的故事。

Arcade Buildings in En'ning Lu

Year Built: Early 1920's
Add: En'ning Lu, Liwan District

Being the best preserved and longest arcade street in Guangzhou, the street links up a great number of historical sites and monuments. Arcade buildings usually consist of shops on the ground floor and residential accommodation upstairs. This building type adapts well to the local conditions and showcases the wisdom of local people.

1—3. 恩宁路的活化，为历史悠久的广州老城区提供了更多的思考，有很好的参考价值。

第五章 | 人居
Chapter 5: Houses

浮华如浮尘，
一代首富的显赫之路

潘家祠街区

树影婆娑，一条路看尽广府与闽南文化融合的建筑艺术，昔日广州十三行的气派，如胶片般逐格回放。

年代 清代
地址 海珠区南华西街

明末清初，广州西关的十三行，客商云集。历史上，广州十三行商人，曾与两淮盐商、山西晋商一起，被称为清代中国三大商人集团，曾是近代以前中国最富有的商人群体。事实上，在广州十三行历史上最显赫的家族，当数以潘振承为首的潘家。

潘、卢、伍、叶四大家族，将眼光越过珠江，选中了一江之隔的"河南"南华西，建别墅，起豪宅。如今，在海珠区的南华西街还保存着潘氏建于1826年的居所，以及一街之隔的古老祠堂。潘家祠、潘氏大院和潘家大院，这是广州近代十三行历史的重要代表建筑。潘家祠保留了部分闽南建筑特征，是广府与闽南地域文化融合在建筑艺术史上的典型代表。

| 1 | 2 | 3 |

1.潘家祠现存建筑只是当年三路五进建筑的一小部分，大部分已被拆除或改建了。南华西街、同福路仍保留着体量完整的骑楼群。
2.西关大屋典型的趟栊门仍在。
3.屋内花阶砖。

The Block of Former Pan Clan Ancestral Hall

Year Built: Qing Dynasty
Add: Nanhua Xijie, Haizhu District

The combination of Guangzhou and southern Fujian architectural styles can be seen in this area. The old days of the former Thirteen Factories, a neighbourhood of warehouses and stores for foreign trade, can still be traced in this area.

1—3.街道空间。

风雨满楼，
春光满路

文明路骑楼

女儿墙内见乾坤，南洋风格骑楼建筑在这条路上烙下了岁月的印记。

年代 20世纪初
地址 越秀区文明路

广州骑楼从20世纪初开始出现，形式多样、保存完整，它既结合了中西传统文化的特点，又体现了广州包容开放的城市精神。骑楼结合西方建筑与岭南建筑的特色，借用柱廊空间，一来可以避风雨、防日晒，二来便于敞开铺面、陈列商品以招徕顾客。

广东地区很早就有下南洋打工谋生的传统，这种传统反映在骑楼建筑上也十分常见，在文明路上的不少骑楼都会在女儿墙上开一个或者多个圆形或者其他形状的洞口，其原本是为了预防南洋一带强烈的台风袭击，减少对建筑物的风负荷的技术处理，从而形成了南洋地区独特的建筑艺术形态。

繁华的往事或不可追寻，留在文明路上的只有时光流逝烙下的深深印记，以及青砖灰瓦记载的岁月变迁。

Arcade Buildings in Wenming Lu

Year Built: Early 1920's
Add: Wenming Lu, Yuexiu District

The Southeast Asian style arcade buildings along this road have not withered in the long history. With the ground floor opening up to a public arcade, they shelter people from wind, rain and sun and provide spaces for business owners to display goods and attract customers.

画风清奇的道场，
康有为幼时曾在此读书

小蓬仙馆

亭台山石，曲径清幽，这座不可多得的精妙建筑，因康有为之名得以整体保存，是古建整体搬迁的范本。

年代 清代
地址 荔湾区芳村大道中 275 号醉观公园内

小蓬仙馆建于清咸丰年间，为石、砖、木穿斗式梁架三进建筑，坐北向南，有大殿、精舍，殿后有小花园，曲径清幽，亭台山石，风景秀丽，至今仍可窥其建筑水平之高。

前殿正面石额"小蓬仙馆"，苍劲有力，两座侧门的门楣上方书有"镜清"和"砥平"字样，山墙檐下一排烘托，上有"花开富贵"图，均为砖刻，线条清晰，刻工精细。头门正面墙体用水磨青砖丝缝砌筑，檐下一块长约10米、宽40厘米的长幅木刻浮雕，雕有花果蔬菜图案20余组，极富岭南地方特色，是不可多得的木刻艺术品。

1898年戊戌政变以后，康有为遭通缉而流亡海外，他在广州的万木草堂被清政府查没。辛亥革命后，国民政府以小蓬仙馆补偿给康有为。

2001年，由于兴建珠江隧道需要，小蓬仙馆得以保护性整体迁建到醉观公园，在玉兰、古松掩映间，这座岭南传统特色古建风貌依然。

1	3	4
2		

1. 原来的小蓬仙馆门口便是通向珠江的小河涌，今天整体搬至醉观公园，倒也清幽。
2. 麻石门拱券石匾，左书"镜清"、右书"砥平"，每块石匾四边皆有精美砖雕框楣衬托。
3—4. 若不是小蓬仙馆一度为康有为家族产业，以叶名琛的名声，断保不住这处修仙的道场，以仅存的构件不难看出，建筑本身，还是非常清秀怡人的。

Small Pengxian Hall

Year Built: Qing Dynasty
Add: Inside Zuiguan Park, 275, Fangcun Dadao Zhong, Liwan District

Being a stone-brick-wood structure of column and tie construction, the Small Pengxian Hall has three halls and two enclosed courtyards. The garden attached to the Hall has a picturesque view with elaborately designed pavilions, terraces and rockery, showcasing the ingenuity and expertise of the builders. It is said that Kang Youwei, a scholar and reformer of the late Qing Dynasty, once studied here when he was young.

第五章｜人居
Chapter 5: Houses

每一座花园，
都有写不完的故事

新河浦建筑群

这里是广州城市现代转型历程中特定区域即特定时期居住建筑发展的产物。洋人兴之，华侨盛之，而革命时期的独特属性，又使东山多了一抹亮色。

年代 近代
地址 越秀区新河浦、龟岗一带

　　老广心中的旧时光，如果说西关是富贵喧闹的烟火气，那么东山就是雅致静谧的洋花瓶。从新河浦路顺着河涌走，阳光透过绿荫洒在春园的红砖柱子上，斑驳的苔痕在光影间若隐若现，穿过中共三大旧址，进逵园看个艺术展，然后缓缓爬上培正中学旁的斜坡路，简园和明园分立在路的两侧，再绕道东山堂听听唱诗班的彩排，或者继续深入龟岗看看昔日华侨富商或文人政客的旧居……这是东山洋房旧建筑群的游览地图，也是不少新广州人的周末休闲自拍路线。

Xinhepu Building Clusters

Year Built: In modern times
Add: Xinhepu and Guigang area, Yuexiu District

Located in the former Dongshan District, the building clusters in Xinhepu were firstly initiated by foreigners and became popular among the returned overseas Chinese. Some of them were later used for revolutionary purposes. Guess who once lived in these garden houses?

1	2	3	4
5	6	7	8

1.中共三大会址纪念馆。
2—3.东山百年老洋房中的五大名园之一逵园。
4.东山百年老洋房中的五大名园之一的简园。
5.基督教东山堂。
6.东山百年老洋房中的五大名园之一的明园。
7.蚁美厚故居。
8.秦牧故居。

165

第五章 | 人居
Chapter 5: Houses

1	2	3	7	8
4	5	6		

1—6.红砖墙、罗马柱、花阶砖、柚木扶手、铁艺栏杆，这些美好的小洋房，正变身成为一栋栋艺术馆和网红小馆，焕发新的生命。
7—8.今天，花木掩映的东山洋房，仍令人百看不厌。

Overseas Chinese returning from Europe and America combined the architectural styles from the countries they used to live and the local architectural style of Lingnan region when building the houses, including the most famous five overseas Chinese houses: the Spring Garden, Jian Garden, Kui Garden, Ming Garden and Yu Garden. Nestled among flowers and trees, they are just too beautiful to be overlooked.

广州城东地形高低不平，清末以前，东山一带尚属偏僻。清光绪末年，外国传教士在东山购地建房，教堂、书局、学校先后落成，至1911年广九铁路通车，外国人和华侨富商在东山大建屋宅、投资房产，东山始兴。

红砖墙、古典柱的两三层独立式别墅，通风采光皆好的花园式洋房在东山龟岗、新河浦一带陆续出现，欧美华侨将侨居地的建筑特色与岭南建筑风格融合，"五大侨园"——春园、简园、逵园、明园、隅园等现存的中西合璧小洋楼，便是当时所建。

岁月变革，风起云涌，这些屋宅或成就过历史大事，或住过叱咤风云的名人，建筑空间与人、事、物丝丝相扣。昔日春园面朝小河，旁有慈姑塘，阡陌交通，蕉林环绕，举目空旷。于田野间建起一式三栋三层砖石混凝土结构的西式小洋楼，屋主气派尽显，加之券廊、爱奥尼柱、新旧交融的围墙，俨然创业有成的华侨富商架势。今天，宁静祥和的小洋房迎来朝圣历史遗迹的人潮。1923年中共三大在广州召开，中共中央机关办公便设在新河浦路的春园，会议期间，陈独秀、李大钊、毛泽东和共产国际代表马林等领导人就住在春园中幢二楼，当年的"国际"餐厅如今复原在纪念馆内，更添画面感。

中共三大旧址，早已旧貌换新颜。而对面美国华侨马灼文建造的逵园，今则被活化成艺术空间，经营着年轻一代的梦想。水泥花阶砖、木楼梯扶手、柚木门窗和铁艺窗花都还在，乍看不出岁月的痕迹，唯有屋顶小山花上镶嵌着建造年代的灰塑提醒着人们，逵园已近百岁。

167

Garden Residences for Returned Overseas Chinese

Year Built: Since 1950's
Add: The Overseas Chinese New Village, Huanshi Donglu, Yuexiu District

A village of garden residences was created for the returned overseas Chinese in Guangzhou after the founding of the New China in 1949. The tranquil courtyards are designed to follow the undulating terrain and nestle among the flowering trees. Thanks to the concerted efforts of architects and planners, the Overseas Chinese New Village still presents a timeless beauty even today.

花荫树影，处处燕归来

华侨新村

新中国成立后，广州为海外归侨建造了一片花影掩映、庭院幽静、山势起伏的花园村落，集众多建筑师、规划师之力营造的华侨新村，至今散发耐看之美、岁月之韵。

年代 20世纪50年代起
地址 越秀区环市东路华侨新村

1	3	4
2		

1. 拥有更多景观空间的半圆形大露台是华侨新村的庭院建筑很常见的设计语言，而当年规划的立体绿化，正深深浅浅地拥抱着这片村落。
2. 漏窗、凉台、树的浓荫……都给人居带来了极大的舒适度。
3. 这栋由广州设计院设计的大宅，是整个华侨新村唯一的中式建筑。
4. 粤剧名伶红线女的故居，现在亦向游客开放。

新中国成立后气象万新，海外侨民回国参与新中国建设的热情高涨，广东是侨民最多的省份，在省会广州为归侨建立宜居之所，自是当时要务。

所选之地，正有可借地势的三处山岗，以此为骨架，有了今天高低错落、步步生景的华侨新村。新村的顾问团队可谓当时建筑界的星宿海：林克明为技术指导，设计委员会成员列名的有佘畯南、黄适、金泽光、陈伯齐、余清江……林克明、金泽光自法国而归，陈伯齐自德国而归，黄适自美国而归，佘畯南是越南归侨，参与规划设计的何运廷是马来西亚归侨，参与建筑设计的朱奇康是缅甸归侨……

这一拨海归，自然十分理解归侨的居住习惯：庭院式住宅、立体绿化、花园式新村，所以华侨新村以独院式住宅建筑为主。而中华人民共和国成立初期百废待兴，林克明倡导的"以艺术的简洁和实用的价值，写出最高之美"、陈伯齐提出的"建筑的实际功能也就是以实用性为出发点来设计其外在形式"，在这种以实用性为出发点的设计思路指导下，这一带的建筑，以简朴整洁为美，建筑整体轻盈明快。

植被设计上立体而多元，西向广植遮阴良好、花香芬芳的白兰花树，干道种植高大蔽日的南洋楹；同时根据广州日照长、雨水多的特点，多采用窗檐滴水、琉璃漏窗、宽大凉台、深邃门廊、屋顶花园等手段，增加居住的舒适度。华侨新村建成之后，印尼归侨知名人士许崇德、加拿大侨领梁葆常、归侨散文大师秦牧、岭南建筑泰斗莫伯治、东南亚归侨、粤剧表演艺术家红线女等名人相继搬进。

今天华侨新村花植浓密，花影婆娑之间洋房、庭院随地势布列，散发着岁月之美。

第六章：新风
Chapter 6: New Trends

广州福地人才济济，近代一批现代岭南派建筑师，接受了西方现代教育，吸收了国外建筑的风格与技术，开风气之先，为这座包容并蓄的城市，因地制宜地开创出更加多样化的优秀建筑。

Guangzhou is blessed with abundant talents. A group of architects of the modern Lingnan School, who were educated in the west with foreign architectural style and technology, became the trend-setter and created more diversified and site-specific buildings for this inclusive city.

莫大师打造的广州名片，
藏而不露，缩龙成寸

Mountain Villa Hotel
山庄旅舍

第六章 | 新风
Chapter 6: New Trends

莫大师打造的广州名片，
藏而不露，缩龙成寸

山庄旅舍

岭南派建筑大师莫伯治，尤善借景造景，山庄旅舍当属杰作，是跨越传统与现代的岭南庭院建筑典范。

年代 1965
地址 白云区白云山风景区内

1	3	4
2		

1. 山庄旅舍的门廊，借坡地而造，四周密植山茶、丹桂、广玉兰和禾雀花，沿坡而上，未及进门，身心已陶然。
2. 两排住房内以水池相隔，空间添了生韵，又自成清凉小气候。
3. 两侧游廊，起伏有致，既是变化的景观动线，又能使整体建筑结构轻盈明快。
4. 每处房间，都各有小景，每处空间皆充满了惊喜。

Mountain Villa Hotel

Year Built: 1965
Add: Inside Baiyun Mountain Scenic Area, Baiyun District

As a city icon designed by Mo Bozhi, master architect of Lingnan School, the hotel features scattered buildings at various heights amidst garden sceneries, showcasing Lingnan-style courtyard architecture that spans over tradition and modernity.

　　山庄旅舍，这座结合了传统造园艺术的现代岭南庭园建筑，一步一景，是岭南派建筑大师莫伯治的杰作。其总体建筑布局因山就势，随地形起伏，建筑围绕不同标高的庭园分散布局，依"前坪—中庭—内庭—后院"的庭园空间序列展开，以现代的建筑形式衬托传统的山池树石，巧于因借，与环境结合得自然而贴切。

　　山庄旅舍尤其以蛇形游廊、中庭环廊、三叠泉等设计最为精妙。1965年5月，周恩来总理、陈毅副总理与印度尼西亚第一副总理兼外长苏班德里约在此举行会谈，这是山庄旅舍建成后接待的第一次国际会议。

　　如今游园，建筑高低错落，庭院绿荫掩映，是城中难得的一处清幽地。

第六章 | 新风
Chapter 6: New Trends

1	3
2	4

1. 闹市之间，一墙之隔的兰圃，经数十年之功，早已树木葱茏。
2. 兰香满园、园势起伏、水系盘匝，小小兰圃，可赏之处甚多。
3. 入门小径，先抑后扬，更显庭院深幽。
4. 平地造景，却是处处变化、步步见景。

在烦嚣深处，
有这处花木清幽品茶地

兰圃 /

闹市绿洲，诗意清幽，小中见大，赏兰品茶，意境无穷。

年代 20 世纪 50 年代
地址 越秀区解放北路 901 号

与越秀山遥望相对，有一座寻幽探胜之地——兰圃。兰圃曾接待过邓小平、陶铸、荣毅仁等国家领导人和尼克松、西哈努克、李光耀等国际贵宾，朱德更将自己培植的兰花赠予兰圃并在此题词留念。其中芳华园是中国参加慕尼黑国际园艺展的中国庭园缩景，以占地少、景点多而闻名于世，被评为"最佳庭园"，获两项金质奖。

几代园林艺术家将兰圃建成一个植物生态园林，栽有200多个品种、近万盆兰花。布局上因地制宜，纵横序列，起承转合，东区建筑具岭南风，西区则吸收了京华园林的堂皇富丽和江南园林的玲珑剔透。空间对比收放自如，亭台楼阁和园内空间各自独立又互相渗透。进园探幽，植物与墙体巧妙阻隔，空间化直为曲，步移景异。小桥流水杜鹃山半隐半现，移步春光亭则豁然开朗，沿阶而下钓鱼台，亭下俯首观鱼水榭，抬头观北面汉白玉宝塔与云天，意境无穷。

闹市里有此绿洲，含蓄隐秀，诗意芬芳，供市民赏兰品茶，诗意里接着烟火气。

Lanpu (Orchid Garden)

Year Built: 1950's
Add: 901, Jiefang Beilu, Yuexiu District

An oasis in a bustling urban setting, the Orchid Garden is a botanic wonderland where one can enjoy both the Lingnan-style architecture and the garden that presents a beautiful mix of Beijing and Jiangnan styles. Ever looking for a secluded place to enjoy orchids and tea in the city? This is it.

第六章 | 新风
Chapter 6: New Trends

于小处见天地，岭南建筑派的造景典范

北园酒家

1958年，莫伯治设计的北园酒家建成。那一年梁思成先生在粤，有人问他最喜欢广州哪座建筑，梁先生毫不迟疑地回答："北园酒家。"

年代 近代—现代
地址 越秀区小北路202号

广州饮食业繁盛，茶楼酒家是岭南地方文化的一个特殊的记号。北园酒家是广州三大园林酒家之一，由建筑师莫伯治先生在原云泉山馆的废址上重新设计建造。

被梁思成列为最欣赏之广州新建筑的北园酒家，建筑与园林环境融为一体，又有浓郁的地方风格，著名画家刘海粟亦为北园写下"其味无穷"的背书。自此，莫伯治声名大噪，泮溪、南园、白天鹅等酒家园林佳作不断。

北园酒家空间布局紧凑，深远曲折的综合式内院，南北两部分以漏花云墙门洞作为过渡，中央挖池造景，绕池分设斋轩亭，以木桥连接。每一座建筑都设开敞式客座，与内院池景相连接，享美食亦赏美景。

莫老带团队多次到珠三角各地收集、拯救了一大批险做薪柴的门窗、家具和木雕部件，这些民间工艺建筑旧料被运回广州加工，为北园酒家带来了跨越时光的别样风致。莫老亦开创了将废弃的旧雕饰、窗扇、屏风和门扇等构件重新运用到建筑里的先例，又将岭南庭园中的山石、水、植物等要素相结合，既体现了地方色彩，又降低了造价。

那些飞罩、落地罩、花罩、隔扇、屏门、满洲窗……岭南庭院建筑中常见的装饰形式，就这样原汁原味地保留、又重现于新的建筑与空间中，在承接传统之际，又有新的表达，称得上是中国造园的重大突破。

生动的山、水、花，端丽的亭、廊、轩、馆、厅堂，内外空间互通渗透，人坐于厅堂却仿如浸润在大自然中。正如明代文震亨所指，对室庐设计，要令居之者忘老，寓之者忘归，游之者忘倦。而北园之美令人忘老忘归忘倦。

Beiyuan Restaurant

Year Built: In modern times
Add: 202, Xiaobei Lu, Yuexiu District

In addition to the exquisite morning tea culture, the North Garden Restaurant, its architecture and gardens are equally celebrated. While maintaining the tradition, the designer has new creations of his own. Therefore, the architect becomes a master of Lingnan school of architecture.

1	2	3
	4	

1. 不过2000平方米的北园酒家，却处处见景。
2. 草木郁葱、水景灵动，美食入喉，美景入眼。
3. 处处可见景，时时亦可入画。
4. 满洲窗。

岭南派风格融入，现代建筑的经典范例

白云宾馆

白云宾馆是高低层结合的庭院式旅馆，高层为客房，低层部分为公共服务设施，利用原有地形构成不同大小的室内庭院联系餐厅和楼梯间的中庭。

年代 1976
地址 越秀区环市东路367号

Baiyun Hotel

Year Built: 1976
Add: 367, Huanshi Donglu, Yuexiu District

As a courtyard hotel including both high-rise and low-rise buildings, the hotel features interior gardens of varied size that follow the original terrain and natural landscape. The atrium connecting the restaurant and the staircase is a highlight in design.

　　1972年，为扩大对外贸易，适应交易会的需要，中央决定在广州兴建白云宾馆。1976年6月1日，这座新中国第一座最高的现代建筑正式开业，无论其建筑风格、结构技术在当时均处于全国领先地位，充分体现了岭南派建筑风格的特色。

　　著名的岭南建筑大师莫伯治有一句名言："没有自然界配合的建筑，起码不是一个完美的建筑。"白云宾馆的设计，就很好地体现了他的这一思想。白云宾馆是高低层结合的庭院式旅馆，高层为客房，低层部分为公共服务设施，利用原有地形构成不同大小的室内庭院，设计的成功之处在于联系餐厅和楼梯间的中庭。中庭利用原有的三棵古榕作为景点，四周的瀑布、景石、水池的组合形成了一个丰富变换的空间，使中庭古拙中带着典雅，简朴中带着丰富，建筑师采用了岭南庭院造园手法进行了大胆的探索，建成后轰动全国，并于2017年12月，入选第二批中国20世纪建筑遗产。

　　白云宾馆是岭南派风格融入现代建筑的一个典范，它的设计水平得到同行的公认，在当时居于全国前列。现在看来，它仍然显得很新潮、很现代化。可以这么说，最有特点的东西就是最现代化的东西，与众不同，就永远无所谓过时、不过时的问题。

Making Gardens like a Painter

Year Built:
Red Bridge in the Liuhua Park: At the end of 1950's
Liuhua West Garden: 1964
Zigzag Bridge in Dongshan Lake Park: At the end of 1950's

These gardens and site furnishings, which were built since 1950's and 1960's when the local citizens were mobolized to excavate the artificial lakes, present a simple yet lasting beauty. Many of these attractive views with which the local citizens grow up have been included into the city's list of historical buildings and will be cherished as masterpieces by the generations to come.

第六章｜新风
Chapter 6: New Trends

以作画之心造园，
画一城山色湖光

流花湖公园红桥、流花西苑｜东山湖公园九曲桥

这些二十世纪五六十年代广州全民动员开挖人工湖公园时代开始建造的园林小景，有着百看不厌的朴素之美，今天，许多伴随着广州人成长的美筑美景，已入选广州历史建筑，成为传世的经典。

年代 流花湖公园红桥：20 世纪 50 年代末
　　　流花西苑：1964 年
　　　东山湖公园九曲桥：20 世纪 50 年代末

地址 流花湖公园：越秀区东风西路 100 号
　　　东山湖公园：越秀区东湖路 123 号

从1958年开始，为解多雨的广州城山洪和内涝之苦，广州全民动员，开挖麓湖公园、东山湖公园、荔湾湖公园、流花湖公园四大人工湖。

虽都是水域造景的公园，得益于一拨从传统文化中成长又受西方文明熏陶的建筑师、规划师、园林设计师一众合力，这四大公园，在设计上在成本上克制又克俭，一山一湖一堤一桥一丘一径一树一花，简朴大方又呈现出各自迥异的风貌，显现出设计师们的深厚功底。

像麓湖公园则借白云山余脉成"半岭隐涛声"，将以往伤人性命的白云山山洪以洼地储蓄，再连通东濠涌导出珠江；而荔湾湖公园则复刻出"两岸荔枝红"的岭南水乡风貌；流花湖公园以热带风情见长——笔直一条蒲葵郁葱的长堤，中间则是园林大师郑祖良设计的红桥，堤岸生葵风、红桥映虹影，这一组"虹影葵风"

和展示盆景的"流花西苑"一样，现在同为广州历史建筑，流花西苑建成后由岭南盆景创始人孔泰初先生担任盆景技术指导。孔先生是广州盆景艺术研究会的总干事，广州成立专业的盆景研究会比上海还早6年，海内外皆闻名。

东山湖公园则以桥的营造见长，其九曲桥、落虹桥、拱桥、五孔桥、三曲桥将水面有序连接，整个空间充满了变幻韵律，而当中的九曲桥，亦入选广州历史建筑，成为见证广州变化的经典之作。

在那一时期，一大批才华横溢的设计师，不再囿于建广厦盖高阁，而是兢兢业业以土地为画板，为公众造园景作美画，留下了充满生命力、带有浓厚岭南色彩的公共建筑和园林小品，时间虽又向前五六十年，这些朴素端庄的作品宛若扎根生在广州的山水间，生在广州人的生命里。

2	3
1	

1. 那一批建筑师出身的郑祖良等人出手设计园林景观，充满结构美，亦注重了环境的构筑与呼应。
2. 流花西苑当时代表着中国盆景的最高水准，空间设计上也表现出处处有画意。
3. 九曲桥使东山湖水域充满了灵动之美。

作者名录
Author List

本书收录了众多机构和摄影师提供的精彩图片，在此表示感谢。图片版权归属于图片提供者。

This book contains wonderful pictures provided by many institutions and photographers, to whom we express our thanks. The copyright of the images belongs to the providers of the images.

大写岭南：
P2 P4 P5 P8 P10 P12 P13 P16 P17 P18 P19 P20 P24
P25 P26 P27 P28 P30 P31 P32 P33 P36 P38 P39 P40
P41 P42 P43 P44 P45 P46 P48 P49 P50 P51 P56 P57
P58 P59 P60 P61 P62 P63 P64 P67 P68 P70 P71 P72
P73 P75 P78 P79 P80 P82 P84 P85 P86 P88 P89 P90
P91 P94 P96 P97 P100 P101 P102 P103 P104 P105 P106
P107 P108 P109 P110 P111 P112 P113 P114 P115 P116
P117 P118 P120 P121 P122 P123 P124 P125 P126 P127
P128 P129 P130 P131 P132 P133 P136 P137 P138 P139
P140 P141 P142 P143 P144 P146 P147 P148 P149 P152
P153 P155 P156 P157 P158 P159 P160 P161 P162 P163
P164 P165 P166 P167 P168 P169 P172 P174 P175 P176
P177 P178 P179 P180 P181 P182 P183

大写岭南（插画）：
P4 P5 P6 P7 P8 P9 P10 P11 P6 P7 P14 P15 P52
P53 P98 P99 P134 P135 P150 P151 P170 P171

广州市城市规划设计公司：
P34 P66 P77 P81~83 P54

胡建军：
P23 P54 P92

梁洁红：
P74

摄影作者：
陈文杰   陈 欣   耳东尘   胡建军   黄得珊
黄庆衡   李 波   李卫东   梁洁红   梁宇星
刘朝宽   刘晓明   龙建平   罗宜威   吕凤霄
孟俊锋   欧阳永康 彭庆凯   饶国兴   史丹妮
孙 兰   覃光辉   王玉龙   温建红   吴术球
叶秉新   张秀珍   邹庆辉

文字作者：史丹妮   郑 宇   孙海刚   崔玛莉
         张 远   莫尔多姿

插画作者：高 毅   龙志放   陈丁财

翻　译：梁 玲

广州建筑图册
广州新建筑图册

NEW ARCHITECTURE

一域商都新韵

广州市规划和自然资源局　编

SPM 南方传媒 | 花城出版社
中国·广州

图书在版编目（CIP）数据

广州建筑图册. 广州新建筑图册 / 广州市规划和自然资源局编. -- 广州：花城出版社，2023.12
ISBN 978-7-5749-0098-1

Ⅰ. ①广… Ⅱ. ①广… Ⅲ. ①建筑文化－广州－图集 Ⅳ. ①TU-092

中国国家版本馆CIP数据核字(2023)第254900号

主编单位：广州市规划和自然资源局
承编单位：广州市城市规划设计有限公司
　　　　　广州市设计院集团有限公司

出 版 人：张 懿
责任编辑：陈诗泳
责任校对：卢凯婷
技术编辑：林佳莹
装帧设计：广州市耳文广告有限责任公司

书　　名	广州建筑图册·广州新建筑图册 GUANGZHOU JIANZHU TUCE · GUANGZHOU XIN JIANZHU TUCE
出版发行	花城出版社 （广州市环市东路水荫路11号）
经　　销	全国新华书店
印　　刷	佛山市迎高彩印有限公司 （佛山市顺德区陈村镇广隆工业区兴业七路9号）
开　　本	787毫米×1092毫米　16开
印　　张	11　1插页
字　　数	108,000字
版　　次	2023年12月第1版　2023年12月第1次印刷
定　　价	288.00元（全三册）

如发现印装质量问题，请直接与印刷厂联系调换。
购书热线：020 - 37604658　37602954
花城出版社网站：http://www.fcph.com.cn

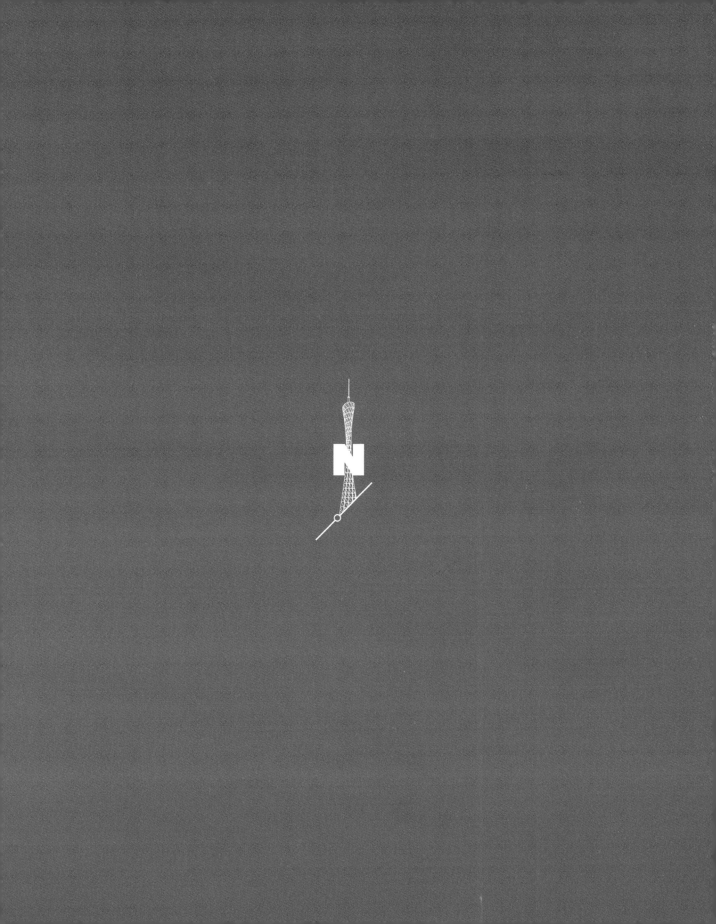

在广州塔顶部的高度俯览珠江两岸
Canton Tower

广州，城市新轴线
Guangzhou, New Central Axis

远眺广州珠江新城夜景天际线
Guangzhou, Zhujiang New Town

前言
Preface

广州新筑迹成长40年
4 Decades of the Urban Development of Guangzhou

广州，一座因水而生，因水而兴，因水而变的城市，一座集历史文化名城、改革开放先锋和现代国际都会于一身的城市。

广州拥有2200多年的历史。唐朝中期，官府开始在民间推广使用秦砖汉瓦技术，广州迎来了第一次城市蜕变，奠定了以北京路为中轴线的广州传统城市的空间格局，城市的重要建筑、商肆也在这一带云集。

20世纪20年代，广州吸取了西方最先进的城市规划建设思想，构建了近代城市空间架构，开始了广州城市的第二次蜕变。这一时期城市重心西移，起义路成为广州近代的纪念性轴线，聚集了众多的纪念性建筑物和公共开放空间。

1978年，改革开放的春风吹拂广州，广州市场经济日渐成熟，城市活力大规模迸发，形式多样。市井商贸繁荣，娱乐消费市场平民化，先进的技术和文化不断进入，灵活、实用、兼容、竞争、创新等现代意识不断加强。到80年代末，广州的建筑事业得到空前的发展，白天鹅宾馆等一批中外合作酒店名扬中外，南越王墓博物馆、天河体育中心等，也是这时期的代表作。90年代，由于经济的发展，大型商业中心和超高层建筑也获得了很大的发展，如天河城、中信广场等。

城市迅猛发展的同时，也带来了生态环境方面的挑战。21世纪元年，广州城市总体战略规划出台，明确了"南拓北优东进西联"的空间战略方针，山城田海的空间格局，三纵四横的生态廊道，明确了用地、交通和生态三大要素是未来城市发展的重要主题。广州由此迎来了第三次蜕变。按战略规划，广州重新打造了城市新中轴线，北起白云山南麓燕岭公园，沿广州东站、天河体育中心、花城广场、海心沙、广州塔、岭南广场、湿地公园，直至珠江后航道中的海心沙岛，全线长12公里。

2000—2010年这10年里，广州以重点建设项目推动战略规划实施，如奥体中心、大剧院、会展中心、省博物馆、西塔、"小蛮腰"、图书馆等；通过新项目的选址实现战略规划的蓝图，如大学城、亚运城等；极力维护城市生态安全格局，如白云山、万亩果园的保护措施等。历经这10年的改造和建设，广州面貌焕然一新，2010年的广州亚运会，让世界看到了一个现代化的国际大都市，广州的国际知名度和综合竞争力大幅提升。

亚运之后，广州继续保持健康发展的步调，越来越多的办公楼与商业中心在珠江新城拔地而起，设计精巧华美的建筑物和舒适怡人的市民广场为羊城塑造出全新的城市形象。"四馆组团"和"三塔夹江"的空间布局最终形成。而南沙自贸区的成立，是后亚运时代的新起点，是广州发展的新契机和新挑战。2017年《财富》论坛举办、2018年世界航线发展大会、2019年世界港口大会、2020年世界大都市协会等高端国际会议接踵而至。经历一次次的蜕变，未来，广州继续寻找着与资源环境共生共存的城市发展之路。

改革开放的40多年，是广州城市建设步伐加快的40多年。这40多年的巨变，揭示了这座千年商都长盛不衰的密码——开放引领、创新不止，向世界诠释了"中国奇迹"背后的时代真理——"开放带来进步，封闭必然落后"。

这本书是广州改革开放40多年现代建筑发展的一个缩影，让我们一起回顾和感受这40多年间广州的**新建筑之美**。

前言
Preface

Guangzhou, a city born by a river, has also been nourished and transformed by the river. It is not only known as a pioneer city of China's Reform and Opening-up, but also famous as a modern metropolis retaining cultural and historical heritage.

During the mid-Tang Dynasty (618-907), the government started to introduce the techniques of making bricks and tiles in the style of the Qin and Han Dynasties (202 BC-220) to the people, which marked the city's first transformation, laying the spatial pattern of Guangzhou with Beijing Road as its central axis. Important buildings and businesses of the city also gathered along with it.

In the 1920s, having absorbed the most advanced western urban planning and construction ideas, Guangzhou initiated its second transformation by constructing a modern urban spatial structure. During this period, the center of the city moved westward, and Qiyi Road became the commemorative axis of Guangzhou in modern times, with a large number of memorial buildings and public open spaces clustering in this area.

In 1978, as the spring breeze of the Reform and Opening-up has blown over Guangzhou, its market economy matured day by day. As a result, the city unleashed unprecedented vigor in various forms, which boosted the development of commerce and trade as well as the popularity of entertainment consumption market. The advanced technology and culture have been continuously introduced to the city, and the modern awareness of flexibility, practicality, compatibility, competition and innovation has been increasingly strengthened. By the end of the 1980s, Guangzhou's construction industry had been enormously developed. A series of hotels eligible for accommodating foreign visitors, such as the White Swan Hotel, became well-known at home and abroad. The Archaeological Site Museum of Nanyue Palace and Tianhe Sports Center were also the masterpieces of this period. In the 1990s, thanks to further economic development, large-scale commercial centers and super high-rise buildings have also burgeoned, such as Teemall and CITIC Plaza.

The rapid development of cities, however, also poses challenges to the ecological environment. In the first year of the 21st century, the general strategic plan of Guangzhou was introduced, which defined the spatial development strategy of "southward expansion, northward optimization, eastward extension and westward combination". The spatial pattern of mountains, cities, fields and seas, and the "three vertical and four horizontal" ecological corridors were also proposed in the report. Land use, transportation and ecology were enshrined as three major elements of future urban development. Guangzhou thus ushered in the third transformation. According to the strategic plan, Guangzhou has rebuilt a new central axis running 12 kilometers, starting from the north at Yanling Park at the south foot of Baiyun Mountain, along Guangzhou East Railway Station, Tianhe Sports Center, Flower City Square, Haixinsha Island, Canton Tower, Lingnan Square and Wetland Park, down to South Haixinsha Island in the lower Pearl River.

From 2000 to 2010, Guangzhou promoted the implementation of the strategic planning with pilot construction projects, such as the Olympic Sports Center, Guangzhou Opera House, Guangzhou International Convention and Exhibition Center, Guangdong Museum, Guangzhou International Finance Centre, Canton Tower and Guangzhou Library. Through the site selection of the new projects including Guangzhou Higher Education Mega Center and Asian Games Town Gymnasium, the city achieved its strategic planning blueprint. It also made enormous efforts to safeguard the urban ecological security via measures of protecting the Baiyun Mountain and the Guangzhou Wanmu Orchard Wetland. After a decade of renovation and construction, Guangzhou has taken on a brand-new look. The 2010 Guangzhou Asian Games have presented Guangzhou as a modern international metropolis to the world. The international popularity and comprehensive competitiveness of Guangzhou have been thereafter greatly enhanced.

After the Asian Games, Guangzhou maintains a sound pace of development. More office buildings and commercial centers are springing up in the Zhujiang New Town. Exquisitely designed buildings and refreshingly comfortable public squares forge a new image of Guangzhou. With the convening of a series of high level meetings, including the 2017 Fortune Forum, China Innovation Conference, China Bioindustry Convention, China Venture Capital Forum, China Innovation and Entrepreneurship Fair, Canton Tower Science&Technology Conference and Guangzhou International Award for Urban Innovation, it marks the final realization of the new central axis. It also symbolizes the formation of the spatial layout of "the Guangzhou Library, Guangdong Museum of Arts, Guangzhou Science Museum and Guangzhou Cultural Center being clustered" and "the Pearl River being Surrounded by the Canton Tower, the East and the West Towers". The establishment of Nansha Free Trade Zone, regarded as a new starting point in the post-Asian Games era, presents new opportunities and challenges for Guangzhou's development. After rounds of transformation, Guangzhou continues to seek the way in which urban development could coexist with the reservation of resources and environment.

During the 40 years of Reform and Opening-up, Guangzhou has accelerated the pace of urban construction. The profound changes in the past 40 years have revealed the secret of the city's everlasting prosperity, namely opening-up and innovation. Moreover, they have also explained to the world the truth of the Chinese Miracle, which is that "opening-up brings progress, while closing-up is bound to lag behind".

This book is an epitome of the development of modern architecture in Guangzhou over the past 4 decades of the Reform and Opening-up, allowing us to review and appreciate the aesthetics of Guangzhou's new architectural achievements over this period of history.

目录
Contents

城脉
City Artery

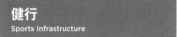

健行
Sports Infrastructure

曜文
Glorious Culture

广州塔	13	天河体育中心	49	西汉南越王博物馆	76
Canton Tower		Tianhe Sports Center		Museum of the Mausoleum of the Nanyue King	
广州国际金融中心（西塔）	17	广东奥体中心	53		
Guangzhou International Finance Center (West Tower)		Guangdong Olympic Sports Center		逸夫人文馆	79
		广州新体育馆	57	Shaw Building of Humanities	
周大福金融中心（东塔）	21	Guangzhou Gymnasium		广东科学中心	82
Chow Tai Fook Financial Center (East Tower)		海心沙亚运公园	61	Guangdong Science Center	
		Haixinsha Asian Games Theme Park		广州大剧院	85
富力盈凯广场	25	广东奥林匹克体育中心游泳跳水馆	64	Guangzhou Opera House	
R&F Yingkai Square		Guangdong Olympic Sports Center Swimming and Diving Hall		广东省博物馆	89
越秀金融大厦	27			Guangdong Museum	
Yuexiu Financial Tower		广州亚运城综合体育馆	66	南越王宫博物馆	92
珠江城大厦（烟草大厦）	31	Asian Games Town Gymnasium		Archaeological Site Museum of Nanyue Palace	
Pearl River Tower（Tobacco Building）		广州国际体育演艺中心	69		
		Guangzhou International Sports Arena		广州新图书馆	95
广晟国际大厦	32			New Guangzhou Library	
The Pinnacle (Guangzhou)		南沙体育馆	71		
利通大厦	37	Nansha Gymnasium			
Leatop Plaza					
中信广场	41				
CITIC Plaza					

6

悦宾
Hospitality

繁津
Downtown Development

会通
Connectivity

白天鹅宾馆 White Swan Hotel	103
中国大酒店 China Hotel	106
花园酒店 The Garden Hotel	108
广州圣丰广场 Guangzhou Shengfeng Plaza	113
W酒店 W Hotel	115

天河城 Teemall	121
正佳广场 Grandview Mall	125
保利国际广场 Poly International Plaza	129
广州太古汇 Taikoo Hui	133
天环广场 Parc Central	137
保利天幕广场（琶洲眼） Poly Skyline Square (Pazhou Eye)	139

琶洲国际会展中心 Guangzhou International Convention and Exhibition Center	145
广州白云国际机场 Guangzhou Baiyun International Airport	149
白云国际会议中心 Guangzhou Baiyun International Convention Center	152
广州南站 Guangzhou South Railway Station	155
广州城市规划展览馆 Guangzhou Urban Planning Exhibition Center	159

城脉
City Artery

在改革开放的背景下，广州经济迅速发展，城市空间结构发生了巨大的变化，一个新的商业及商务中心逐渐在以天河体育中心为核心的天河地区形成。新世纪的规划调整了珠江新城的规划定位，首次提出了建设广州市21世纪城市中央商务区的目标和实施措施，超高层商务办公建筑飞速发展，广州城市新轴线逐步形成。

今天的花城轴线周边规划建设了39幢建筑，它不是一个冰冷的城市CBD，而是荟萃婀娜多姿的广州塔、相辅相生的东西双塔、繁花似锦的花城广场等众多城市地标的城市人文中心，更有引领亚热带绿色建筑技术的珠江城，造型新颖的广晟国际大厦、利通大厦、富力盈凯广场等，它们丰富了广州的天际线，向世人展现广州全新的城市面貌和绿色先进的时代理念。

During Reform and Opening-up, Guangzhou has witnessed rapid economic development with the urban spatial structure undergoing tremendous changes. A new commercial and business center has gradually formed in Tianhe District with Tianhe Sports Center as its core. The planning for the new century adjusted the positioning of the Zhujiang New Town. For the first time, the goal and the implementation measures for the construction of Guangzhou Central Business District in the 21st century were proposed. With the rapid development of super high-rise commercial office buildings, the new central axis gradually came into being.

There are 39 buildings being planned and constructed along today's Huacheng axis. It is not just a business district, but a cultural center with many city landmarks, including the Canton Tower, the East and the West Towers and the Flower City Square. It also houses the Zhujiang New Town, which is at the forefront of subtropical green building technologies, and other well-designed buildings including the Pinnacle, the Leatop Plaza and the R&F Yingkai Square. They have enriched Guangzhou's skyline and shown the world Guangzhou's new image and commitment to sustainability.

1999 年，广州提出了"城市新轴线"的概念，要创造一个空间上集聚的城市中心。经过 20 年城市地标的辉煌建设，一座座人们耳熟能详的新建筑在新轴线两侧拔地而起。

"In 1999, the concept of a 'new central axis' was proposed to create a well-organized city center in Guangzhou by grouping various buildings together. After two decades of construction, numerous new buildings and landmarks, now widely recognized, have sprung up on both sides of this new central axis."

1978

1978 国务院利用外资建设旅游饭店领导小组决定在广州建设五座五星级宾馆
1978 改革开放城市重心逐渐东移
1981 全国第一个超级商场——广州友谊公司超级自选商场开业
1982 广州入选第一批国家级历史文化名城
1983 —1985 三大中外合资宾馆先后建成开业
1984 广州经济技术开发区成立
1982 广州图书馆开馆

1983 白天鹅宾馆 White Swan Hotel
1984 中国大酒店 China Hotel
1984 花园酒店 The Garden Hotel
1987 天河体育中心 Tianhe Sports Center
1988 西汉南越王墓博物馆 Museum of the Mausoleum of the Nanyue King

1988

1987 第六届全国运动会在广州召开
1987 广州规划二沙岛文化建筑
1985 天河客运中心动工兴建
1990 地铁首期工程获批，全长12.7公里
1993 开始建设珠江新城
1994 正佳广场奠基
1996 作为内地最早一批商贸类的天河城开业
1997 广州地铁开通第一条线路
1997 广州公路主枢纽首个大型公用客运站——天河客运站建成
1998 广东奥林匹克体育馆动工兴建

1996 天河城 Teemall
1997 中信广场 CITIC Plaza
1997 广东美术馆 Guangdong Art Gallery

1998

1999 天河城广场全面开业
1999 提出珠江新城轴线规划
2000 启动《广州城市建设总体战略概念规划纲要》，提出"南拓、北优、东进、西联"八字方针
2001 《广州城市建设总体战略概念规划纲要》通过
2001 第九届全国运动会在广州召开
2002 通过并正式实施南沙开发区规划
2004 珠江新城轴线四大公建计划全面启动
2004 广交会正式移师琶洲会展中心
2005 国务院批复《广州市城市总体规划 (2001—2010年)》
2005 正佳广场、太古汇、万菱汇、广州地铁3号线开通
2006 《亚规》修编，在八字方针基础上增加"中调"战略
2007 《广州2020'城市总体发展战略规划》开展

2001 广东奥体中心 Guangdong Olympic Sports Center
2001 广州新体育馆 Guangzhou Gymnasium
2002 琶洲国际会展中心 Guangzhou International Convention and Exhibition Center
2003 逸夫人文馆 Shaw Building of Humanities
2004 广州白云国际机场 Guangzhou Baiyun International Airport
2005 正佳广场 Grandview Mall
2006 保利国际广场 Poly International Plaza
2007 白云国际会议中心 Guangzhou Baiyun International Convention Center
2008 广东科学中心 Guangdong Science Center

2008

2010 广州国际金融中心(西塔) Guangzhou International Finance Center (West Tower)
2010 广州塔 Canton Tower

北京路旧照
An old picture of Beijing Road in Guangzhou

越秀区起义纪念馆，原明代指挥都司署
Guangzhou Uprising Memorial Hall in Yuexiu District, the former Military Government Office in the Ming Dynasty (1368-1644)

规划建设前的珠江新城区域旧照
An old picture of Zhujiang New Town before the planning and construction

建设中的广州塔和东塔
The Canton Tower and East Tower in construction

广州的珠江新城城市新轴线
Guangzhou's new central axis in Zhujiang New Town

2017年的《财富》全球论坛在广州举办
The 2017 Fortune Global Forum was held in Guangzhou

2017 《深化粤港澳合作 推进大湾区建设框架协议》签署
2017 广州《财富》论坛举办
2016 作为天河路商圈收官之作的天环广场正式开业
2015 提出打造三大枢纽、完成一江两岸三带的目标
2014 《广州市历史文化名城保护规划》获批复
2011 国务院正式批复《广州市土地利用总规划（2006—2020年）》
2010 首届亚洲残疾人运动会在广州开幕
2010 广州塔亮灯竣工
2009 粤剧列入世界非物质文化遗产名录

2018

2018 广州城市规划展览馆 Guangzhou Urban Planning Exhibition Center
2017 保利天幕广场（琶洲眼） Poly Skyline Square (Pazhou Eye)
2016 周大福金融中心（东塔） Chow Tai Fook Financial Center (East Tower)
2015 天环广场 Parc Central
2015 越秀金融大厦 Yuexiu Financial Tower
2013 珠江城（烟草大厦） Pearl River Tower (Tobacco Building)
2012 W酒店 W Hotel
2012 广州新图书馆 New Guangzhou Library
2012 富力盈凯广场 R&F Yingkai Square
2012 利通大厦 Leatop Plaza
2011 广州太古汇 Taikoo Hui
2011 广晟国际大厦 The Pinnacle (Guangzhou)
2010 广州南站 Guangzhou South Railway Station
2010 南越王宫博物馆 Nanyue Palace Archaeological Site Museum
2010 广东省博物馆 Guangdong Museum
2010 广州大剧院 Guangzhou Opera House
2010 南沙体育馆 Nansha Gymnasium
2010 广州国际体育演艺中心 Guangzhou International Sports Arena
2010 广州亚运城综合体育馆 Asian Games Town Gymnasium
2010 广东奥体中心游泳跳水馆 Guangdong Olympic Sports Center Swimming and Diving Hall
2010 海心沙亚运公园 Haixinsha Asian Games Theme Park
2010 广州圣丰广场 Guangzhou Shengfeng Plaza

City Artery

广州塔
Canton Tower

地址：广州市海珠区阅江西路222号
建成时间：2010年
建筑高度：塔身主体高454米，天线桅杆高146米，总高度600米
总建筑面积：129 699平方米
占地面积：17.546万平方米

Address: No.222 Yuejiang West Road, Haizhu District, Guangzhou
Completed in: 2010
Height: The main tower body is 454m high. The antenna is 146m high. The total height of the tower is 600m.
Total building area: 129,699 m²
Floor area: 175,460 m²

要说广州最重要的城市名片，民间已经将原来的五羊雕像用广州塔替换。它上下宽，中间细，腰部扭转，高高的塔尖耸入云霄，宛如一位亭亭玉立的少女，又似定海神针旋出美丽舞韵，因这性感的造型，它又被称为"小蛮腰"。但是，真正官方认可"小蛮腰"为广州城市名片应该是在2017年12月广州举办《财富》论坛期间，由著名视觉设计师曹雪教授设计的广州城市新Logo"广州"二字共同勾勒出广州现代建筑地标——广州塔（小蛮腰），这个Logo让全世界记住了广州。

在这座城市景观最重要的制高点上，你可以在423米世界最高的旋转餐厅品味中西美食，在454米世界最高的摩天轮上俯瞰城市美景，在488米世界最高的户外观景平台上欣赏日落，在世界最高最长的空中云梯漫步，挑战485米世界最高的垂直速降游乐项目。广州塔就是这样一座令人称奇的，以观光旅游为主，并有科普教育、文化娱乐和城市窗口功能的城市地标。

广州塔的设计结合了建筑、结构、雕塑和美学，扭腰、偏心的造型使"小蛮腰"在不同的方向看都显现出不同的形态。复杂的平面功能、超高、扭转、偏心、透空、收腰、消防疏散、抗风抗震等都是前所未有的设计和建造难题。广州塔的建成，是飞速发展中的广州向世界骄傲地递出一张新名片。

城脉
City Artery

The Canton Tower has replaced the previous Five Goat Statue as the most important landmark in Guangzhou. It is wide up and down and narrow in the middle with a twisting structure and has a soaring spire, which makes it look like a fair lady or the twisted "Magic Cudgel" in Chinese mythology. This sexy shape earns it a nickname as "Slim Waist".

At the most important vantage point of the city's landscape, you can savor both Chinese and western cuisine at a revolving restaurant in the height of 423 meters, have a bird-view of the city in the highest Ferris wheel at 454 meters height, appreciate sunset at the highest outdoor observation deck at a height of 488 meters, have a stroll on the longest open-air skywalk, or experience the highest vertical free fall at 485 meters height.

城脉
City Artery

广州国际金融中心（西塔）
Guangzhou International Finance Center (West Tower)

地址：广州市天河区珠江西路5号　　Address: No.5 Zhujiang West Road, Tianhe District, Guangzhou
建成时间：2010 年　　　　　　　　Completed in: 2010
建筑高度：437.5米　　　　　　　　Height: 437.5m
层数：地上103层，地下4层　　　　Floors: 103 floors above ground, 4 floors below ground
总建筑面积：45.6万平方米　　　　　Total building area: 456,000 m²
占地面积：3.1万平方米　　　　　　Floor area: 31,000 m²

在广州珠江新城的规划中，最重要的超高层建筑非东、西双塔莫属。在1993年的珠江新城规划国际竞赛中，来自美国的托马斯夫人的胜出方案中就有两栋350米的建筑放在黄埔大道边，这就是东塔和西塔的前身。1999年将这两栋超高双塔移至南面珠江边，矗立于文化中心区之后，目的是在珠江新城轴线和珠江的交会处形成戏剧性的汇集点，与珠江对岸的广州塔形成"三塔夹江"的城市景观。

2004年，广州国际金融中心（西塔）动工，建成后的西塔是全球十大超高层建筑之一。西塔是全幕墙结构，采用双层呼吸式幕墙系统，能产生通透的效果，当初的设计立意为"江畔水晶"。西塔的主塔楼平面是圆弧三角形，建筑结构采用独特创新的巨型斜交网格支撑体系，呈现出清晰可见的流动几何图案，"钻石形状斜网格"结构随着高度逐步进行尺寸缩减，使得楼体给人一种"直入云霄"的视觉震撼。整个建筑外立面精美流畅，典雅现代，极为晶莹剔透而又瑰丽多彩，犹如一块通透的水晶。

2011年，广州国际金融中心获得CTBUH（世界高层建筑与都市人居学会）颁发的"最佳高层建筑大奖"；2017年，经美国绿色建筑委员会考评，西塔凭借国内最高分的成绩总分88分正式获得能源与环境设计先锋评级（LEED）V4铂金级认证，是国内第一个以LEED V4标准获得运营阶段铂金级认证的超高层地标建筑；2019年，获得BOMA（国际建筑业主与管理者协会）中国COE（人力资源管理专家中心）认证，成为华南地区首座具有国际运营管理体系的超高层综合体地标建筑，是亚洲首个获得BOMA认证的REITS（不动产投资信托基金）资产项目；2020年，CTBUH授予广州国际金融中心全球"十年奖"。

一域商都新韵
New Architecture

In the planning of the Zhujiang New Town, the most important super high-rise buildings are the East Tower and the West Tower. The twin towers, situated on the southern part of the central axis, along with the Canton Tower on the opposite side, form a landscape of the Pearl River being surrounded by the three towers. The construction of the West Tower started in 2004. After its completion, the Tower became one of the ten global super high-rise buildings, consisting of office building, five-star hotel, shopping mall, apartment and international conference center. The main tower is characterized by a rounded triangular plan and a spindle-like shape sketched out by an oblique grid. The hidden framing glass exterior wall creates a kind of exquisite, elegant, and modern effect, which makes the tower a crystal-like building standing on the Flower City Square.

周大福金融中心（东塔）
Chow Tai Fook Financial Center (East Tower)

地址：广州市天河区珠江东路6号
建成时间：2016年
占地面积：26494平方米
总建筑面积：507681平方米
建筑高度：530米
层数：地上111层，地下5层

Address: No.6 Zhujiang East Road, Tianhe District, Guangzhou
Completed in: 2016
Floor area: 26,494 m^2
Total building area: 507,681 m^2
Height: 530 m
Floors: 111 floors above ground, 5 floors below ground

  2009年，号称"广州之巅"的东塔奠基。它是集超五星级酒店及餐饮、服务式公寓、甲级办公楼、地下商城等功能于一身的综合性超高层建筑。建成后的东塔建筑以超过西塔100米的高度一跃成为广州第一高塔。这座号称珠江新城的"收官之作"至此完美落幕。

  东塔采用刚柔并重的外形，以直线条为主面，建筑外形轮廓分明。塔身采用独特的"之"字形退台设计，划分功能的同时，还在高度上呼应了相邻建筑的高度。东塔的结构首次采用超高强绿色混凝土，减轻东塔自重的同时还增加了使用面积。它的玻璃幕墙采用白色陶土板挂件，可以天然采光，打造出如同直冲云霄的水晶体般迷人的整体效果。另外，值得一提的是世界上最快的电梯在广州周大福金融中心（东塔）落户，该电梯的最快速度达到1260米/分钟（秒速21米）。

After the completion of the West Tower, the East Tower marked the realization of the construction of Zhujiang New Town, which attracted much expectation. It reaches a height of 530 meters, surpassing the West Tower by 100 meters, which makes it the tallest tower in Guangzhou. The exterior is rigid but smooth. The main facade featuring straight lines allows the outline of the building to be well defined. The twin towers accommodate buildings in different heights and shapes, which implies the openness and inclusiveness of Guangzhou. The West Tower looks more gentle, while the East Tower is more rigid. They complement each other and enliven the dynamic skyline along the central axis.

富力盈凯广场
R&F Yingkai Square

地址：广州市天河区华夏路16号	Address: No. 16 Huaxia Road, Tianhe District, Guangzhou
建成时间：2012年	Completed in: 2012
占地面积：7942平方米	Floor area: 7,942 m^2
总建筑面积：17.45万平方米	Total building area: 174,500 m^2
建筑高度：296.5米	Height: 296.5 m
层数：地上66层，地下5层	Floors: 66 floors above ground, 5 floors below ground

　　富力盈凯广场是集商场、5A甲级办公楼、服务式公寓、酒店于一身的大型综合体。它的外形简洁而独树一帜，交错内切的角落与玻璃幕墙上参差的竖向金属板幕墙让人联想起当地特有的竹子，从下往上看又犹如丝丝雨滴倾斜而下，恍如在重力的影响下，垂直而下的"雨滴"。宽度和密集程度随着高度的上升逐渐减小，使建筑从视觉上变得更为扎实和稳固。

　　富力盈凯广场正方形的体量尊重了街区网格的硬朗几何线条，退让出面向城市的公共空间。多样化的建筑功能和与城市交通系统的连接让富力盈凯广场迅速成为城市中不可或缺的一员，并为这座城市不断变化的天际线增添了一抹独特的色彩。

The rectilinear shape of the R&F Yingkai Square respects the geometric rigidity of the street grid and provides citizens with extra open spaces between itself and its neighboring buildings. The Square promptly stands out as an indispensable part of the city for its mixed-use functions and the connection to the city transportation system, adding a touch of uniqueness to the ever changing skyline.

越秀金融大厦
Yuexiu Financial Tower

地址：广州市天河区珠江东路28号	Address: No.28 Zhujiang East Road, Tianhe District, Guangzhou
建成时间：2015年	Completed in: 2015
占地面积：10836平方米	Floor area: 10,836 m²
总建筑面积：21万平方米	Total building area: 210,000 m²
建筑高度：390.4米	Height: 390.4 m
层数：地上68层，地下4层	Floors: 68 floors above ground, 4 floors below ground

  越秀金融大厦定位是"国际办公楼新起点"，率先提出"健康办公"理念——让健康融入工作，让工作回归自然，采用高效的冷热源设备、全热回收新风系统、太阳能光热技术、可调节外遮阳、雨水回收利用和光导管照明技术等，为进驻的企业商户提供新鲜的空气和健康的工作环境，成为中轴线上的"绿洲"，越秀金融大厦开启了广州办公楼领域的新里程。

  2018年，经美国绿色建筑委员会综合考评，越秀金融大厦凭借总分97分——全球最高分的显赫战绩一举夺得有绿色建筑界"奥斯卡"之称的LEED 当中"建筑运营与维护（EBOM）"类别V4铂金级认证。

  越秀金融大厦的落成不但为天河CBD的整体商业形象与品牌价值提升增添更多的砝码，还为这一中国南方商业心脏地带注入可持续发展的创新驱动力，将超甲级绿色建筑引领至新的高度。

The Yuexiu Financial Building is positioned as a "new starting point for international office buildings" and is the first to propose the philosophy of providing a healthy office environment. The core idea of this philosophy is allowing people to work in a healthy and natural environment. With the adoption of high and new technologies, the building has become the "oasis" on the central axis by providing people with fresh air and a healthy working environment. The green concept has been well implemented from its design to construction. In 2018, by 97 marks, the highest mark in the world, the Yuexiu Financial Building was awarded the LEED EBOM Platinum certification, known as the green buildings Oscar. This allows it to become an icon of the green buildings in Guangzhou.

珠江城大厦（烟草大厦）
Pearl River Tower (Tobacco Building)

地址：广州市天河区珠江西路15号	Address: No.15 Zhujiang West Road, Tianhe District, Guangzhou
建成时间：2013年	Completed in: 2013
占地面积：1万平方米	Floor area: 10,000 m²
总建筑面积：21万平方米	Total building area: 210,000 m²
建筑高度：309米	Height: 309 m
层数：地上71层，地下5层	Floors: 71 floors above ground, 5 floors below ground

珠江城大厦曾被国外媒体誉为"世界最节能环保的摩天大厦"，一经问世就蜚声国内外。

大厦的设计者之一SOM合伙人Richard Tomlinson介绍说：珠江城大厦有几个有意思的地方，首先它的朝向设计为南偏东13°，可以使大楼获得更多的风能和太阳能；其次建筑物立面上有几个漏斗状的洞穴，这个构思是源自结构工程学，它可以迫使风从洞穴通过大楼，可以减少风对大楼的冲击，使整个大楼减小摆动、更为稳定，通过这样的设计，达到了节省钢和混凝土的目的；此外，还在漏斗形的通风口增设了风力涡轮发电机，可作为额外的电力供应。

大厦以"零能耗"为最高设计目标，采用了十一项绿色节能技术以及可再生能源利用技术，节能率达63.2%。2013年，广州珠江城大厦荣获美国绿色建筑委员会LEED铂金级绿色建筑认证。珠江城项目对建筑节能的推广，带动了国内一些开发商更多地进行建筑节能技术的探索，对推动中国建筑节能市场做出了卓越贡献。

The Pearl River Tower has been called by foreign media as the most energy-efficient skyscraper in the world. It has attracted much attention and respect after its completion. Designed with its original goal of a "zero-energy" building, the Pearl River Tower incorporates 11 energy-saving and renewable energy technologies, with an energy-saving rate of 63.2%. In 2013, the Tower passed the LEED Platinum certification developed by the U.S. Green Building Council.

一域商都新韵
New Architecture

广晟国际大厦
The Pinnacle (Guangzhou)

地址：广州市天河区珠江西路17号
建成时间：2011年
占地面积：7907平方米
总建筑面积：15.6万平方米
建筑高度：350米
层数：地上60层，地下6层

Address: No.17 Zhujiang West Road, Tianhe District, Guangzhou
Completed in: 2011
Floor area: 7,907 m^2
Total building area: 156,000 m^2
Height: 350 m
Floors: 60 floors above ground, 6 floors below ground

广晟国际大厦整体形象新颖独特，韵律感极强，因为造型酷似一支巨型铅笔，民间俗称——"铅笔大楼"。在高层鳞次栉比的珠江新城CBD区，广晟国际大厦外立面广泛采用了石材幕墙。建筑的顶部通过退台变化以及不同材质的对比来丰富层次，加上立面以竖向线条为构图母题，建筑造型显得挺拔而富有韵律感，稳重中散发新锐气息。挺拔的造型，石材、金属和玻璃组成的外墙，诠释出高贵的新古典建筑风格，在珠江新城CBD现代建筑群中格外与众不同。

大厦的设计遵循环保理念，适当控制幕墙面积，并通过先进的构造技术避免旧式安装工艺中由填缝工序带来的污染。外墙采用LOW-E双层中空玻璃、浅色石材和低采暖系数材料，有效降低建筑能耗。

The Pinnacle is innovative and vibrant in its overall look. Its resemblance to the shape of a giant pencil makes it known as the "Pencil Building". The change of the set-back model on the top and the contrast of different materials have enriched the design of the building. Furthermore, the graphic pattern on the facade adopts vertical lines as its theme, which makes the building look tall and vibrant and displays a sense of steadiness and innovation. The upright shape and the exterior wall composed of stone, metal and glass demonstrate the neoclassical architectural style, which allows it to stand out from its neighbors in Zhujiang New Town.

城脉
City Artery

利通大厦
Leatop Plaza

地址：广州市天河区珠江东路32号
建成时间：2012年
占地面积：9916平方米
总建筑面积：16万平方米
建筑高度：303米
层数：地上65层，地下5层

Address: No.32 Zhujiang East Road, Tianhe District, Guangzhou
Completed in: 2012
Floor area: 9,916 m^2
Total building area: 160,000 m^2
Height: 303 m
Floors: 65 floors above ground, 5 floors below ground

  利通大厦是珠江新城中轴线北大门上收口的一个关键项目，具有强烈标志性。它是一座超甲级办公楼，通过关注摩天大楼设计最根本的几个问题，采用极简主义设计，实现了令人叹服的标志性效果。建筑风格简洁、通透、线条清晰、气质高贵，37.8米高的大堂气势雄伟，简洁精致的建筑主体造型、个性鲜明的斜坡屋面、极具光影表现力的鳞片状玻璃幕墙营造出艺术雕塑的效果，犹如一块晶莹剔透的冰晶体矗立于新城市中轴线的门户。

  利通大厦采用了十大先进节能环保措施，是广州首个获得LEED双认证的超甲级写字楼。

The Leatop Plaza, a super-grade office building, is located at the intersection of Huangpu Avenue and Zhujiang Avenue. Even though it adopts a minimalist design, it has achieved an iconic effect. It is simple, transparent, linear and noble in architectural style. The magnificent 37.8-meter-high lobby, the exquisite geometric-shape main body, the distinct pitched roof and the scaly glass curtain wall with good reflection of light and shadow make the sculpture-like building glittering as crystal.

The Leatop Plaza adopts ten advanced energy-saving measures and received the U.S. LEED Gold Pre-certification in 2011. This is also the first green building in Guangzhou receiving this certification.

城脉
City Artery

中信广场
CITIC Plaza

地址：广州市天河区天河北路233号
建成时间：1997年
占地面积：2.3万平方米
总建筑面积：32.2万平方米
建筑高度：391米
层数：主楼80层，附楼38层，地下2层

Address: No.233 Tianhe North Road, Tianhe District, Guangzhou
Completed in: 1997
Floor area: 23,000 m^2
Total building area: 322,000 m^2
Height: 391 m
Floors: 80 floors of the main building; 38 floors of the residential buildings attached to the main building; 2 floors below ground.

　　竣工于1997年的中信广场，是当时全亚洲最高的楼宇，也是世界上最高的纯钢筋混凝土结构办公楼。它不但让20世纪末的广州拥有了自己的形象名片，更是广州迈向现代化大都市的成熟标志。它见证了广州改革开放以来经济腾飞，见证了商业地产的跌宕起伏，承载了许许多多的梦想与辉煌。很多人对这座霸居广州第一高楼十余年的建筑印象深刻，因为在任何一张介绍广州城市形象的照片里，都可以见到从各种角度仰视的中信广场，"在中信上班"，也一度成为市民们足够有面子的事。

　　中信广场的主楼80层，高达391米，38层的副楼在它的两翼展开，如同一只展翅高飞的雄鹰。"有别于一般华丽的地标建筑，中信广场以其粗犷简朴的外形，展现了高速发展中的城市活力和力量，也反映出发展商崇尚平实简洁的自我风格。"香港的刘荣广伍振民建筑师事务所有限公司介绍当初的设计理念时指出，中信广场整体强调对称，希望传达出城市坚实稳固的前景。

城脉
City Artery

Completed in 1997, the CITIC Plaza was the tallest building in Asia and the tallest reinforced concrete building in the world. It not only was an image card of Guangzhou at the end of the 20th century, but also marked the maturity of Guangzhou towards a modern metropolis. Many people are deeply impressed by this building which held the title of the tallest building in Guangzhou for more than a decade. In any pictures that introduce the image of Guangzhou, you can see the CITIC Plaza taken from various angles. The completion of CITIC Plaza has propelled the construction of the office buildings and the residential buildings around it, which together formed the Tianhe North Business District with CITIC Plaza as its center. Today, CITIC Plaza has become an important landmark and scenic spot at the north end of the central axis.

健行

Sports Infrastructure

要说现代广州城市的发展，离不开这几次体育盛会——1987年的第六次全国运动会、2001年的第九次全国运动会和2010年的亚运会。

1987年在广州举行的第六届全运会，为广州体育建筑的建设，写下了开创时代的篇章。借着六运会的东风，天河体育中心的各个场馆相继落成，包括体育场、体育馆、游泳馆三大建筑，以及各训练场馆和新闻中心等多个配套设施，"六运会"催生了天河体育中心，变出了一个新"天河"，昔日的城郊荒地成为新城市中轴线上的璀璨亮点。

2001年，羊城再次迎来了第九届全运会。这个广州新千年第一个规模盛大的全国综合性体育盛会，为广州增添了两颗明珠：广东奥林匹克体育中心和广州新体育馆，变出了一个"三年一中变"的新广州，山更绿了，水更清了，交通更顺畅了。

时至2010年，第十六届亚运会开幕式的圣火在海心沙的水舞台上方燃起，揭开了广州体育史上最辉煌的一幕。自亚运城行动计划启动以来，广州相继开工建设了八大实施工程，同时，在已有场馆的基础上增添了一系列的国际化体育场馆，如广东奥林匹克体育中心游泳跳水馆、广州国际体育演艺中心等。借此盛会，让世界了解了广州，也让广州走向了世界。在举办开闭幕式的海心沙、健儿们挥洒汗水的南沙体育馆、奥体中心游泳跳水馆等，都让世界记住了广州的辉煌时刻，也体现了广州富有活力的一面，向世人展示了广州市民开放包容、平稳务实的性格。

The modern development of Guangzhou cannot be achieved without these sports events: the 6th National Games of the People's Republic of China in 1987, the 9th National Games of the People's Republic of China in 2001 and the Asian Games in 2010.

The 6th National Games, held in Guangzhou in 1987, ushered in an era of the sports buildings in Guangzhou. After the convening of the 6th National Games, various venues in the Tianhe Sports Center were completed one after another, including the stadium, the gymnasium, the swimming pool, the training venues, the news center and other supporting facilities. The 6th National Games led to the birth of the Tianhe sports Center. The new Tianhe district thus became a reality. The former suburban wasteland has become a shining icon on the new central axis.

In 2001, the 9th National Games was once again held in Guangzhou. It was the first large-scale national comprehensive sports event held in the new Millennium in Guangzhou. The Guangdong Olympic Sports Center and the Guangzhou Gymnasium were therefore built, which witnessed the radical changes of Guangzhou. The ecological environment was better and the traffic was lighter.

By 2010, the 16th Asian Games' flame was lit on the water-themed stage on Haixinsha Island during the opening ceremony, which marks the most brilliant moment in the sports history of Guangzhou. Since the launch of the Asian Games Town Gymnasium, there have been 8 major projects in construction. At the same time, in addition to the existing venues, a series of international sports venues were added, including the Guangdong Olympic Sports Center Swimming and Diving Hall and the Guangzhou International Sports Arena. Guangzhou took this event as an opportunity to go global while allowing itself to be known by the world. Glorious moments were created on the Haixinsha Island where the opening and the closing ceremonies were held, and at the Nansha Gymnasium and the Guangdong Olympic Sports Center Swimming and Diving Hall where the athletes were striving for success. All of these moments allow Guangzhou to be remembered by the world and reflect the vitality of Guangzhou as well as the open, inclusive, and down-to-earth personality of its citizens.

六运会、九运会、亚运会先后在广州举办。这些全国性乃至国际性的体育盛会，加快了广州城市现代化建设的进程，将广州建设的辉煌成就推向了一个又一个历史的高峰。

The 6th National Games, the 9th National Games and the Asian Games were all convened in Guangzhou. These sports galas, both national and international, have accelerated the modernization development in Guangzhou and achieved historic peaks for urban development.

1978

- 1978 国务院利用外资建设旅游饭店领导小组决定在广州建设3座五星级宾馆
- 1978 改革开放城市重心逐渐东移
- 1981 全国第一个超级商场——广州友谊公司超级自选商场开业
- 1982 广州入选第一批国家级历史文化名城
- 1982 广州图书馆开馆
- 1983 广州经济技术开发区成立
- 1983-1985 三大中外合资宾馆先后建成开业
- 1984 广州经济技术开发区成立

- 1983 白天鹅宾馆 White Swan Hotel
- 1984 中国大酒店 China Hotel
- 1984 花园酒店 The Garden Hotel

1988

- 1987 广东奥林匹克体育馆动工兴建
- 1987 第六届全国运动会在广州召开
- 1985 天河体育中心动工兴建
- 1990 地铁首期工程获批，全长12.6公里
- 1993 开始建设珠江新城
- 1994 正佳广场奠基
- 1996 天河城作为内地最早一批商城的天河城开业
- 1997 广州地铁开通第一条线路
- 1998 广州公路主枢纽首个大型公用客运站——天河客运站建成

- 1987 天河体育中心 Tianhe Sports Center
- 1988 西汉南越王墓博物馆 Museum of the Mausoleum of the Nanyue King
- 1997 广东美术馆 Guangdong Art Gallery
- 1997 中信广场 CITIC Plaza
- 1996 天河城 Teemall

1998

- 1999 天河广场全面开业
- 1999 提出珠江新城中轴线规划
- 2001 第九届全国运动会在广州召开，提出"南拓、北优、东进、西联"八字方针
- 2001 《广州城市建设总体战略概念规划纲要》通过
- 2000 启动《广州市建设总体战略概念规划纲要》
- 2002 广州成功获得2010年亚运会的举办权
- 2004 珠江新城中轴线四大公建计划全面启动
- 2004 广交会正式移师琶洲开发区展中心
- 2005 国务院批复《广州市总体规划（2001—2010年）》
- 2005 正佳广场、太古汇、万菱汇等一系列大型商城相继建成开业
- 2005 广州地铁3号线开通
- 2006 《总规》修编，在八字方针基础上增加"中调"战略
- 2007 《L天2020：城市总体发展战略规划》开展

- 2001 广州新体育馆 Guangzhou Gymnasium
- 2001 广东奥体中心 Guangdong Olympic Sports Center
- 2002 琶洲国际会展中心 Guangzhou International Convention and Exhibition Center
- 2003 粤夫人文馆 Shaw Building of Humanities
- 2004 白云国际机场 Guangzhou Baiyun International Airport
- 2005 汇佳广场 Grandview Mall
- 2006 保利国际广场 Poly International Plaza
- 2007 白云国际会议中心 Guangzhou Baiyun International Convention Center

2008

- 2008 广东科学中心 Guangdong Science Center
- 2010 广州国际金融中心(西塔) Guangzhou International Finance Center (West Tower)
- 2010 广州塔 Canton Tower

1987 年 8 月 30 日天河体育中心正式落成
The Tianhe Sports Center was completed on 30 August 1987

六运会开幕式上表演方队举着会徽进场
The performers holding the emblem are entering the venue at the opening ceremony of the 6th National Games of China

六运会开幕式观众用彩纸拼出羊的图案
Spectators use colored paper to create patterns of sheep at the opening ceremony of the 6th National Games of China

九运会开幕式上的表演
The performance at the opening ceremony of the 9th National Games of China

2010 年亚运会开幕式在珠江新城海心沙举行
The opening ceremony of 2010 Asian Games was held on the Haixinsha Island

亚运会上中国名将勇夺男子 4×100 米接力桂冠
Chinese athletes won the gold medal in the Men's 4×100 m relay of the Asian Games

2010 年广州亚残运会闭幕式文艺表演
The cultural performance at the closing ceremony of 2010 Asian Para Games in Guangzhou

2010 年亚残运会上选手获奖现场
The athletes won awards at the 2010 Asian Para Games

2017 《深化粤港澳合作 推进大湾区建设框架协议》签署
2017 广州《财富》论坛举办
2016 作为天河路商圈收官之作的天环广场正式开业
2015 提出打造三大枢纽、完成一江两岸三带的目标
2014 《广州市历史文化名城保护规划》获批复
2011 国务院正式批复《广州市土地利用总体规划 (2006—2020年)》
2010 首届亚洲残疾人运动会在广州开幕
2010 第16届亚洲运动会在广州开幕
2010 广州塔亮灯
2010 亚运城竣工
2009 粤剧列入世界非物质文化遗产名录

2018 广州城市规划展览馆 Guangzhou Urban Planning Exhibition Center
2017 保利天幕广场（琶洲眼） Poly Skyline Square (Pazhou Eye)
2016 周大福金融中心（东塔） Chow Tai Fook Financial Center (East Tower)
2015 天环广场 Parc Central
2013 珠江城（烟草大厦） Pearl River Tower (Tobacco Building)
2012 越秀金融大厦 Yuexiu Financial Tower
2012 W 酒店 W Hotel
2012 广州新图书馆 New Guangzhou Library
2012 富力盈凯广场 R&F Yingkai Square
2011 利通大厦 Leatop Plaza
2011 广州太古汇 Taikoo Hui
2010 广晟国际大厦 The Pinnacle (Guangzhou)
2010 广州南站 Guangzhou South Railway Station
2010 南越王宫博物馆 Archaeological Site Museum of Nanyue Palace
2010 广东省博物馆 Guangdong Museum
2010 广州大剧院 Guangzhou Opera House
2010 南沙体育馆 Nansha Gymnasium
2010 广州国际体育演艺中心 Guangzhou International Sports Arena
2010 广州亚运城综合体育馆 Asian Games Town Gymnasium
2010 广东奥体中心游泳跳水馆 Guangdong Olympic Sports Center Swimming and Diving Hall
2010 海心沙亚运公园 Haixinsha Asian Games Theme Park
2010 广州圣丰广场 Guangzhou Shengfeng Plaza

2018

健行
Sports Infrastructure

天河体育中心
Tianhe Sports Center

地址：广州市天河区天河路299号
建成时间：1987年
总建筑面积：12.47万平方米
建筑规模：体育场可容纳6万人，体育馆8628个座位，游泳馆3200个座位

Address: No.299 Tianhe Road, Tianhe District, Guangzhou
Completed in: 1987
Total building area: 124,700 m²
Scale: The stadium can accommodate 60,000 people; the gymnasium accommodates 8,628 seats and the natatorium accommodates 3,200 seats.

　　1982年2月，国务院批准广东省承办第六届全国运动会，这是全国运动会第一次在北京、上海以外的地方举行。广州市决定利用原天河机场兴建体育中心，当时还没有天河区，天河体育中心之名就是取自天河机场。

　　当年天河体育中心由一批年轻的建筑师担纲设计，建成体育场、体育馆、游泳馆三大场馆。后来陆续建设了网球学校、棒球场、篮球城、保龄球馆、门球场、亚运体育文化中心等6个场馆。三大场馆在设计上各具特色，均采用了敞开式，以利通风、透气，适应岭南气候条件。大梁、大柱、大形体、大跨度，雄伟壮观，简洁明快，充分体现了体育的速度与力度之美。

　　天河体育中心的建成直接促进了周边区域的开发。30年来，天河体育中心的发展直接带动了天河区的整体发展，其周边逐渐形成闻名全国的天河商圈，也成为广州城市新中轴线上的璀璨明珠。

49

Sports Infrastructure

In February 1982, the Sixth National Games were authorized to be held in Guangdong province by the State Council. Guangzhou, the capital of Guangdong province, decided to build a sports center at the site of former Tianhe Airport. Designed by a group of young architects, the three main stadiums have their own characteristics in design. They all adopt a ventilation-friendly open style to adapt to the climate in Lingnan area. The stadium features large beams, columns, body and span, embodying the spirit of sports, namely speed and strength. Over the past three decades, the development of Tianhe Sports Center has also driven the overall growth of Tianhe District. The Tianhe Commercial Circle, well-known throughout the country, has gradually become a pearl on the new central axis of Guangzhou.

一域商都新韵
New Architecture

健行
Sports Infrastructure

广东奥体中心
Guangdong Olympic Sports Center

地址： 广州市天河区东圃奥体路818号
建成时间： 2001年
总建筑面积： 14.56万平方米
建筑规模： 80012个座位

Address: No.818 Dongpu Olympic Road, Tianhe District, Guangzhou
Completed in: 2001
Total building area: 145,600 m^2
Scale: 80,012 seats

　　广东奥林匹克体育场是目前国内最大、最好，同类场馆中建设速度最快，舞台规模最大，最早采用分开的"缎带"式屋顶的国家级大型体育场。建成至今已成功承办了第九届全运会的开幕式和田径、足球等比赛项目以及2010广州亚残运会的开、闭幕式，成为广州引以为自豪的新城市标志，更被评为羊城新八景之一，名曰"五环晨曦"。

　　这座为九运会全新打造的奥体中心体育场造型新颖、雄伟、浪漫并富有象征意义。它首次打破了国内体育场传统圆形的设计理念，采用了飘带造型的独特设计。体育场内设21个看台区，五颜六色的座椅，犹如万片色彩斑斓的花瓣，汇成广州市的市花——木棉花，极为壮观。
　　广东奥林匹克体育中心除能举办国际最高级别体育赛事外，还可多功能开发经营，构成集竞技体育、群众体育、旅游观光、医疗康复、休闲娱乐于一身的大型体育文化中心。

The Guangdong Olympic Sports Center is the largest, best, and the fastest-completed stadium of its kind in China. The stadium, featuring the largest stage nationwide, is the first state-level large stadium to adopt a separate ribbon-like roof design. Since its completion, it has held the opening ceremony, track and field, football and other matches of the Ninth National Games, as well as the opening and closing ceremonies of the 2010 Asian Games, making it a new landmark that Guangzhou takes pride in.

一域商都新韵
New Architecture

Sports Infrastructure

广州新体育馆
Guangzhou Gymnasium

地址：广州市白云区白云大道南783号
建成时间：2001年
总建筑面积：95748平方米
建筑规模：10018个座位

Address: No.783 South Baiyun Avenue, Baiyun District, Guangzhou
Completed in: 2001
Total building area: 95,748 m^2
Scale: 10,018 seats

广州新体育馆犹如三个形状独特的"叶片"撒落在风景优美的白云山山脚。广州新体育馆由法国著名建筑设计师保罗·安德鲁设计，是一个以体育比赛为主，兼顾文艺表演、会议、展览的多功能综合性体育建筑，涵盖体操、篮球、羽毛球等项目。它于2001年落成，是广州市政府为举办"九运会"而兴建的体育场馆，在这里举办了"九运会"体操比赛和闭幕式。

新体育馆最大的设计亮点是将以人为本的理念与大自然的特色紧密联系到一起，使建筑本身充满着人性化以及自然化的风格，它的三个场馆均采用下沉式设计，大部分建在地下，这样既便于观众的进出交通组织，又能令建筑物置于若隐若现和充满诗情画意的自然之中，与毗邻的优美生态环境融为一体。

一域商都新韵
New Architecture

The Guangzhou Gymnasium looks like three leaves scattered at the foot of Baiyun Mountain.Designed by famous French architect Paul Andreu, the multi-purpose sports architecture, open in 2001, is mainly used to host sporting events, such as gymnastics, basketball and badminton. Cultural performance, conference and exhibition events are also held there. As a gymnasium built by the Guangzhou Municipal Government for the Ninth National Games, it was where the gymnastics competitions and closing ceremony of the Games were held.

健行
Sports Infrastructure

Sports Infrastructure

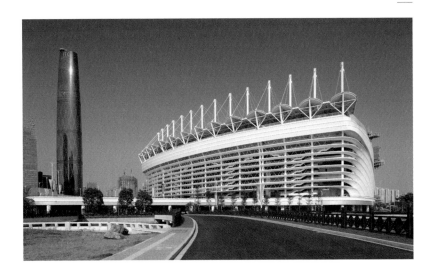

海心沙亚运公园
Haixinsha Asian Games Theme Park

地址：广州市天河区海心沙岛	Address: Haixinsha Island, Tianhe District, Guangzhou
建成时间：2010年	Completed in: 2010
占地面积：17.6万平方米	Floor area: 176,000 m²
建筑规模：可容纳3.5万名观众	Scale: accommodating 35,000 people.

　　在老广州人的眼中，海心沙宁静而神秘，甚至带有少女般的羞涩，静静地躺在珠江之上。海心沙一直被定义为生态和休闲小岛。此前，海心沙岛上只有为数不多的几栋楼房，一座狭窄的小步行桥就是进岛唯一通道。2010年亚运会的到来，把这里改造成了开闭幕式的主场馆。如今，这里已经是南北向新城市中轴线和东西向珠江的交会点。

　　亚运会之所以选择在海心沙岛举办开闭幕式，是因为它独特的地理位置正好符合了开闭幕式场馆要"以珠江为舞台，以城市为背景"理念的要求。以往国际大型体育赛事的开闭幕式，大多选择在封闭的体育场内举行，广州亚运会打破了这一传统，首次从封闭走向开放，将整个海心沙岛打造成一艘扬帆起航的巨轮，实现了时间与空间的延伸与张扬。这种创造性的开闭幕仪式，让整个城市都成为舞台，而此恰恰是广州开放进取城市精神的体现。

一域商都新韵
New Architecture

健行
Sports Infrastructure

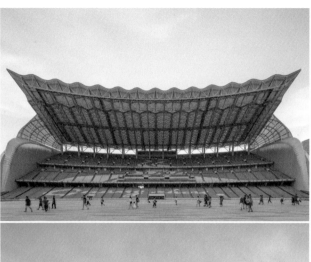

As one of the few small islands of the Pearl River, Haixinsha Island has long been defined as an ecological island for leisure. There were only a few multi-story buildings on the island, and the only access to it was a narrow pedestrian bridge. The 2010 Asian Games transformed the island into the main venue for the opening and closing ceremonies. It used to be that indoor arenas were mostly chosen to hold such ceremonies of international sports events; however, the Guangzhou Asian Games broke the convention, and employed an open-roof stadium instead for the first time. The Haixinsha Island was thus transformed into a colossal ship, realizing the extension of time and space.

广东奥林匹克体育中心游泳跳水馆
Guangdong Olympic Sports Center Swimming and Diving Hall

地址：广州市天河区东圃奥体路818号
建成时间：2010 年
总建筑面积：33331平方米
建筑规模：4584 个座位

Address: No.818 Dongpu Olympic Road, Tianhe District, Guangzhou
Completed in: 2010
Total building area: 33,331 m^2
Scale: 4,584 seats

广东奥林匹克体育中心游泳跳水馆有着广州"水立方"之称，它采用双色螺旋流动造型，40个不同的四边形切面沿着两条不同的方向进行渐变，蓝白相间的切片相互交错，如同生命基础的DNA双螺旋组织，提示着运动与人的关系密不可分。

白色和蓝色相间，既巧妙地隐喻了广州"云山珠水"的城市地理特征，又是对主体育场"飘带"曲线的延续。通过相互穿插流动的造型，结合建筑朝向，很好地满足了建筑内部空间高度、采光通风、建筑节能以及合理布置设备管道的需求。

广东奥体中心游泳跳水馆是2010年广州亚运会暨第16届亚运会游泳跳水项目的主要比赛场馆，是12个新建场馆之一，它承担了第16届亚运会的游泳、跳水和现代五项游泳比赛及亚残运会游泳比赛。

The Guangdong Olympic Sports Center Swimming and Diving Hall is located in the northwest of the Olympic Center and the southwest side of the planned landscape lake. This hall, along with the stadium, the multi-function hall and other buildings, has become important landscape architecture with a demanding view in the Guangdong Olympic Sports Center. It has hosted many competitions during the Asian Games. The architecture adopts a two-color spiral flow model. Forty different quadrilateral sections are gradually morphed along two different directions. The blue and white sections are interlaced with each other, just like the double helix DNA strands, the basis of life, indicating the inseparable relationship between sports and human beings.

广州亚运城综合体育馆
Asian Games Town Gymnasium

地址：广州市番禺区兴亚大道33号	Address: No.33 Xingya Avenue, Panyu District, Guangzhou
建成时间：2010 年	Completed in: 2010
总建筑面积：31482平方米	Total building area: 31,482 m²
建筑规模：6233 个座位	Scale: 6,233 seats

　　广州亚运城综合体育馆位于番禺南部，北临风景优美的莲花湾，与运动员村、升旗广场隔水相望，并通过空中漫步廊道与主新闻中心相连，是进入亚运城的门户。它包括体操馆、台球馆、壁球馆及广州亚运历史博物馆，是2010年广州亚运会比赛的重点场馆之一，同时也是广州市最大的综合性体育馆。

　　体育馆的设计构思凸显了"飘逸彩带"的主题，用流动的线条展现岭南建筑轻灵飘逸的神韵，打造独一无二的建筑风格，建筑造型新颖独特，具有强烈标志性。体育馆的结构复杂多变，使用多种工程新技术和多种节能新技术，减少能耗，绿色环保，屋面雨水可以收集再利用。

　　亚运结束之后，广州亚运城综合体育馆成为广州新城集体育、商业、公共服务等多功能于一身的建筑综合体，服务于市民。

Sports Infrastructure

The Asian Games Town Gymnasium is located in the south of the Asian Games Town, facing the beautiful Lianhuawan to the north. The Gymnasium, including a gymnastics hall, a snooker hall, a squash hall and the Guangzhou Asian Games History Museum, is one of the key venues for the 2010 Asian Games, and is also the largest comprehensive gymnasium in Guangzhou. The theme of flying ribbons is embodied in the design of the Gymnasium. The flowing lines show the light and elegant nature of Lingnan architecture. The innovative architectural shape and style makes the Gymnasium an iconic building.

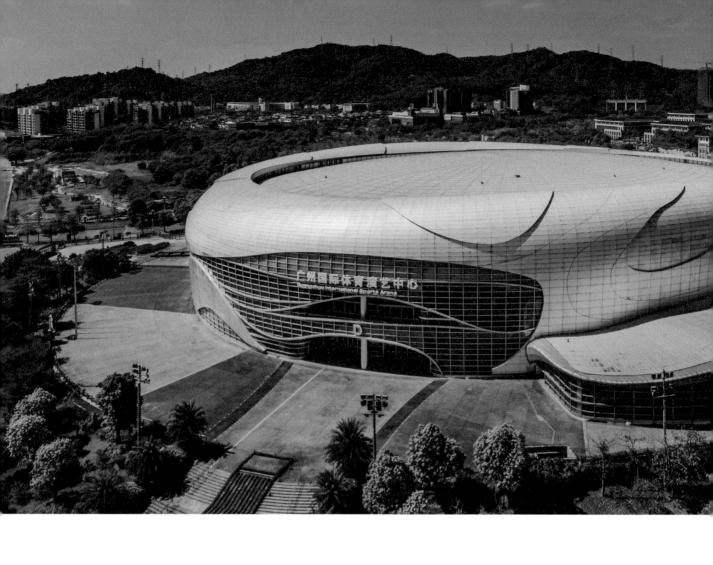

Sports Infrastructure

广州国际体育演艺中心
Guangzhou International Sports Arena

地址：广州市黄埔区大塱村开创大道2666号
建成时间：2010 年
总建筑面积：130312.8平方米
建筑规模：18345 个座位

Address: No.2666 Kaichuang Avenue, Dalang Village, Huangpu District, Guangzhou
Completed in: 2010
Total building area: 130,312.8 m^2
Scale: 18,345 seats

广州国际体育演艺中心是第16届亚运会篮球比赛的主场馆，同时也是亚洲最好的能够举办大型室内演唱会的综合性演艺场馆之一。

广州国际体育演艺中心设计方案由NBA御用设计公司美国Manica建筑设计事务所负责设计，设计理念源自穗城广州美丽的五羊传说，外观整体设计效果既平滑又动感；内部结构设计合理，既能满足亚运会篮球赛、NBA中国赛等大型国际赛事的要求，同时又能举办国际一流的大型演艺节目活动。广州国际体育演艺中心融体育、演艺活动为一体，是创新型的华南体育娱乐建筑。

The Guangzhou International Sports Arena is the main arena for the 16th Asian Games basketball matches. The arena is one of the best multi-purpose performing venues to host large-scale indoor concerts in Asia. With a floor area measuring 130,000 square meters, the arena serves as a new world-class platform for sports and entertainment. Every detail of the arena's design and layout has been taken into careful consideration, making it suitable for hosting world-level events, such as international basketball, ice hockey games and large concerts.

健行
Sports Infrastructure

南沙体育馆
Nansha Gymnasium

地址：广东省广州市南沙区体育三路
建成时间：2010 年
总建筑面积：30236.3平方米
建筑规模：8080个座位

Address: Tiyu 3rd Road, Nansha District, Guangzhou, Guangdong Province
Completed in: 2010
Total building area: 30,236.3 m^2
Scale: 8,080 seats

南沙体育馆作为亚运会武术比赛的主赛场，武术特点与相关议题让设计师不得不关注。建筑平面形态与太极图的巧合，给了设计师巧于利用"武术"概念的机会。设计中，将组成体育馆外壳的九个曲面单元片片层叠，并分为南北两组以比赛大厅圆心为中心呈螺旋放射状展开。运用近似太极阴阳图的构成方式，隐喻中国武术的最高境界——"阴阳俱合，天人合一"。结合广东地区独特的海洋文化特征，借鉴了富有肌理变化的"海螺"外壳作为造型设计的意向，创造出一种蕴含地域文化特征的建筑形态。

南沙体育馆的主体钢结构部分采用了先进的双层环形张弦穹顶结构，主跨度达到了98米，为目前国内此类结构的最大跨度。屋顶设置了大面积的天窗及电动遮阳系统，可调节大厅内光线的强弱，在平时可利用自然采光进行各种活动，节约能源。

The Nansha Stadium is located in Nansha Development Zone at the southernmost part of Guangzhou. It was the venue for Chinese Kungfu and dance sport competitions during the Guangzhou Asian Games in 2010. As one of the new large-scale stadiums built for the Guangzhou Asian Games, it integrates the inclusiveness of Guangdong's marine culture and the innovation in Lingnan architecture. The landscape with immense waves would arouse endless imagination. The designers adopted a composition similar to the Yin-Yang Diagram, suggesting the highest realm of Chinese Kungfu, which is the harmony between Yin and Yang, and between humans and nature.

曜文

Glorious Culture

广州是一个有着2200多年悠久历史的都会城市，自秦汉时期伊始广州便是当时南越国的都城。1983年考古发现西汉初年南越王国第二代王赵眜的陵墓，让广州建都的历史追溯到公元前200年前后。我们现如今仍然能在西汉南越王墓博物馆和南越王宫博物馆中感受这段久远而辉煌的历史。

1982年广州入选了第一批国家级历史文化名城。1987年广州规划于二沙岛建设一批文化建筑，广东美术馆、星海音乐厅等，落成后成为当时全市标志性的文化建筑，为城市文化注入新的活力。

新世纪的广州更加重视文化建设，努力打造岭南文化中心，培育世界文化名城。1999年提出珠江新城轴线规划后，四大公建计划全面启动，全新的文化地标建筑相继面世，有如"明珠江畔的灵石"的广州大剧院、如"宝盒"般的广东省博物馆，以及如"翻开的图书"的广州新图书馆等。

Guangzhou is a city with a history of more than two thousand years. Since the Qin and Han Dynasties (202 B.C.-220), Guangzhou had been the capital of the Nanyue Kingdom. The archaeological discovery in 1983 of the mausoleum of Zhao Mo, the second king of the kingdom in the early Western Han Dynasty, marks the founding of Guangzhou as a capital dating back to around 200 BC. Now we can still experience this long and glorious history in the Museum of the Mausoleum of the Nanyue King and the Archaeological Site Museum of Nanyue Palace of the Western Han Dynasty.

In 1982, Guangzhou was one of the first cities to be listed into the National Famous Historical and Culture Cities. A series of buildings for cultural development were planned on Ersha Island in 1987. Guangdong Museum of Art and Xinghai Concert Hall then became iconic cultural buildings in Guangzhou, injecting new energy into the city's culture.

In the new century, Guangzhou values cultural construction more than ever. In an effort to cultivate a world-class cultural city, the city decided to build the Lingnan Cultural Center. After the Axis Planning of the Zhujiang New Town was put forward in 1999, four major construction plans were launched together in an all-round way, and brand-new cultural landmarks emerged one after another, such as the Guangzhou Opera House by the Pearl River, the treasure-box-like Guangdong Museum, and the Guangzhou Library which looks like an opened book.

站在新时代的坐标上，广州浸润着改革、开放、包容的精神气质，用国际化的视野，通过文化地标的打造，为千年的城市文脉续写绚丽新篇。

Entering the new era, Guangzhou is full of the spirit of reform, openness and inclusiveness. By building cultural landmarks from an international view, Guangzhou starts a new chapter in its long and glorious cultural history.

1978

- 1978 国务院利用外资建设旅游饭店领导小组决定在广州建设5座五星级宾馆
- 1978 改革开放城市重心逐渐东移
- 1981 全国第一个超级商场——广州友谊公司超级自选商场开业
- 1982 广州入选第一批国家级历史文化名城
- 1982 广州图书馆开馆
- 1983–1985 三大中外合资宾馆先后建成开业
- 1984 广州经济技术开发区成立

1988

- 1987 广州规划二沙岛文化建筑
- 1987 第六届全国运动会在广州召开
- 1985 天河体育中心动工兴建
- 1997 广州公路主枢纽首个大型公用客运站——天河客运站建成
- 1998 广东奥林匹克体育馆动工兴建
- 1996 作为内地最早一批商城的天河城开业
- 1997 广州地铁开通第一条线路
- 1990 地铁首期工程获批，全长12.7公里
- 1993 开始建设珠江新城
- 1994 正佳广场奠基

1998

- 1999 提出珠江新城轴线规划
- 1999 天河城广场全面开业
- 2000 启动《广州城市建设总体战略概念规划纲要》，提出"南拓、北优、东联、西联"八字方针
- 2001 第九届全国运动会在广州召开
- 2001 《广州城市建设总体战略概念规划纲要》通过
- 2002 广州成功获得2010年亚运会的举办权
- 2004 珠江新城轴线四大公建计划全面启动
- 2004 广交会正式移师琶洲会展中心
- 2005 国务院批复《广州市城市规划（2001—2010年）》
- 2005 正佳广场、太古汇、万菱汇等一系列大型商城相继建成开业
- 2005 广州地铁3号线开通
- 2006 《总规》修编，在八字方针基础上增加"中调"战略
- 2007 《广州2020：城市总体发展战略规划》开展

2008

- 2008 广东科学中心 **Guangdong Science Center**
- 2007 白云国际会议中心 **Guangzhou Baiyun International Convention Center**
- 2006 保利国际广场 **Poly International Plaza**
- 2005 正佳广场 **Grandview Mall**
- 2004 广州白云国际机场 **Guangzhou Baiyun International Airport**
- 2003 逸夫人文馆 **Shaw Building of Humanities**
- 2002 琶洲国际会展中心 **Guangzhou International Convention and Exhibition Center**
- 2001 广州新体育馆 **Guangzhou Gymnasium**
- 2001 广东奥体中心 **Guangdong Olympic Sports Center**
- 1997 广东美术馆 **Guangdong Art Gallery**
- 1997 中信广场 **CITIC Plaza**
- 1996 天河城 **Teemall**
- 1988 西汉南越王墓博物馆 **Museum of the Mausoleum of the Nanyue King**
- 1987 天河体育中心 **Tianhe Sports Center**
- 1984 花园酒店 **The Garden Hotel**
- 1984 中国大酒店 **China Hotel**
- 1983 白天鹅宾馆 **White Swan Hotel**
- 2010 广州国际金融中心（西塔）**Guangzhou International Finance Center (West Tower)**
- 2010 广州塔 **Canton Tower**

开发建设前的二沙岛航拍图
An aerial picture of the Ersha Island before it was developed

二沙岛上重点规划的星海音乐厅
The Xinghai Concert Hall, one of the key projects on the Ersha Island

二沙岛上重点规划的广东美术馆
The Guangdong Museum of Art, one of the key projects on the Ersha Island

广州大剧院前的演奏
Performance in front of the Guangzhou Opera House

珠江新城轴线上的四大文化建筑
The four cultural buildings on the new central axis in the Zhujiang New Town

南越宫苑遗址发掘现场
The excavating site of the Nanyue Place

庆祝粤剧申遗成功专场演出
A performance celebrating that Cantonese opera being officially inscribed onto UNESCO Representative List of the Intangible Cultural Heritage of Humanity

南越王墓遗址发掘现场
The excavating site of the Mausoleum of the Nanyue King

2017 《深化粤港澳合作 推进大湾区建设框架协议》签署
2017 广州《财富》论坛举办
2016 作为天河路商圈收官之作的天环广场正式开业
2015 提出打造三大枢纽、完成一江两岸三带的目标
2014 《广州市历史文化名城保护规划》获批复
2011 国务院正式批复《广州市土地利用总规划（2006—2020年）》
2010 首届亚洲残疾人运动会在广州闭幕
2010 亚运城竣工
2010 第16届亚洲运动会在广州开幕
2010 广州塔亮灯
2009 粤剧列入世界非物质文化遗产名录

2018 广州城市规划展览馆
Guangzhou Urban Planning Exhibition Center
2017 保利天幕广场（琶洲眼）
Poly Skyline Square (Pazhou Eye)
2016 周大福金融中心（东塔）
Chow Tai Fook Financial Center (East Tower)
2015 天环广场
Parc Central
2015 越秀金融大厦
Yuexiu Financial Tower
2013 珠江城（烟草大厦）
Pearl River Tower (Tobacco Building)
2012 W酒店
W Hotel
2012 广州新图书馆
New Guangzhou Library
2012 富力盈凯广场
R&F Yingkai Square
2012 利通大厦
Leatop Plaza
2011 广州太古汇
Taikoo Hui
2011 广晟国际大厦
The Pinnacle (Guangzhou)
2010 广州南站
Guangzhou South Railway Station
2010 南越王宫博物馆
Archaeological Site Museum of Nanyue Palace
2010 广东省博物馆
Guangdong Museum
2010 广州大剧院
Guangzhou Opera House
2010 南沙体育馆
Nansha Gymnasium
2010 广州国际体育演艺中心
Guangzhou International Sports Arena
2010 广州亚运城综合体育馆
Asian Games Town Gymnasium
2010 广东奥林匹克中心游泳跳水馆
Guangdong Olympic Sports Center Swimming and Diving Hall
2010 海心沙亚运公园
Hanxinsha Asian Games Theme Park
2010 广州圣丰广场
Guangzhou Shengfeng Plaza

2018

西汉南越王博物馆
Museum of the Mausoleum of the Nanyue King

地址：广州市解放北路867号
建成时间：1988年
总建筑面积：17.4万平方米

Address: No.867 Jiefang North Road, Guangzhou
Completed in: 1988
Total building area: 17,400 m²

  1983年夏天，一群凿石刨土的民工在广州市区一个叫象岗的小山包，无意中发现了一块断裂的石板和一个洞穴，从此揭开了尘封千年的南越国王陵的神秘踪迹。

  南越文王墓的出土，集中反映了2000多年前岭南政治、经济和文化等多方面的内容，是迄今为止，岭南地区发现年代最早、规模最大、陪葬物最丰富的汉初古墓，也是岭南地区最早的一座大型彩绘石室墓。发掘后，墓室就地保护，并在其旁边辟建了西汉南越王博物馆。

  1988年正式对外开放的西汉南越王博物馆，主要展示南越王墓原址及其出土文物。博物馆以古墓为中心，依山而建，拾级而上，将综合陈列大楼、古墓保护区、主题陈列大楼几个不同序列的空间有机地联系在一起，突出了遗址博物馆的群体气派，是岭南现代建筑的一个辉煌代表，曾获得6项国内外建筑大奖。

In the summer of 1983, a group of migrant workers, during the process of construction, accidentally found a broken stone slate and a cave under the hill named Xianggang in downtown Guangzhou, which unveiled the mystery of the mausoleum of the Nanyue King. This mausoleum is so far the oldest and largest tomb of the beginning of the Han Dynasty (202 BC–220 AD) with the most funerary objects, and also the oldest large-scale stone-chamber tomb carved with colorful murals in Lingnan area. The chamber was preserved after excavation, and the Museum of the Mausoleum of the Nanyue King was thereafter built next to it.

曜文
Glorious Culture

逸夫人文馆
Shaw Building of Humanities

地址：华南理工大学校内中心区
建成时间：2003年
占地面积：8805平方米
总建筑面积：6398平方米

Address: The central area of the South China University of Technology
Completed in: 2003
Floor area: 8,805 m²
Total building area: 6,398 m²

　　新旧建筑并存、中西文化兼容、人与自然有机融合，是华南理工大学的鲜明特色。 位于校园东、西湖之间的逸夫人文馆就是当中最具有代表性的现代岭南建筑之一。

　　华南理工大学建筑设计研究院的建筑师这样诠释他们的建筑理念：延续以院落为中心的空间格局，保留人们心中"庭院深深"，"绿树成荫"的美好记忆；创造空灵通透，典雅端庄的新岭南建筑。建筑师确定了"少一些、空一些、透一些、低一些"的设计思想，以实现这一特殊场所中人、自然与建筑的共生。

　　充满人文气息的人文馆将人们对基地的记忆和院落文脉延续下去。建筑充分尊重现有院落空间的布局，完全保留基地内的树木，以新的建筑形式围合、限定，并通过空间渗透将三个气氛不同的院落空间联系起来，成为一个整体。人们可以通过廊道、桥梁，从东、南、北三个方向穿行，建立起人与自然、建筑间的对话关系。

The campus of the South China University of Technology features the coexistence of new and old buildings, the accommodation of both Chinese and Western cultures and the integration of people and nature. Located between the east and west lakes of the campus, the Shaw Building of Humanities is one of the most symbolic modern Lingnan buildings of the university. The design retains the courtyard-centered spatial arrangement, allowing people to enjoy the scene of courtyard and tree shade and putting up a new Lingnan building of clarity and elegance. The designer adopted the concept of designing a simpler, lower, more airy and transparent building, to achieve the harmony among people, nature and the building.

一域商都新韵
New Architecture

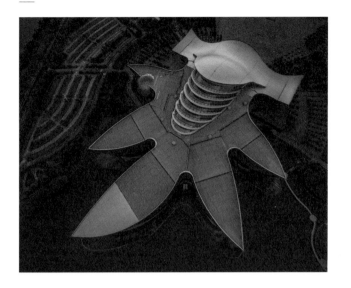

广东科学中心
Guangdong Science Center

地址：广州大学城西六路168号
建成时间：2008年
占地面积：45万平方米
总建筑面积：14万平方米

Address: No.168 West 6th Road, Higher Education Mega Center, Guangzhou
Completed in: 2008
Floor area: 450,000 m²
Total building area: 140,000 m²

广东科学中心位于名校云集的广州大学城小谷围岛，是亚洲规模最大的科普教育基地之一。它的主体建筑以"科技航母"为造型，象征广东科技的发展：探索、追求、一往无前，又像是大学城凝聚智慧结晶盛放开来的一朵"科学之花"，向人们展示着这里的风采。在设计理念上，表现了科技先锋、乘风破浪、灵动人文和吉祥如意四大寓意。

广东科学中心建筑面积共有14万平方米，内有八大主题展馆、4座科技影院、两个开放实验室和一个2500平方米的"数字家庭体验馆"，此外还有建筑面积达2万平方米的临时展区，用于随时展示国内外最新的科学技术成果。广东科学中心从正面看，像一只灵动的科学"发现之眼"，它以"自然、人类、科学、文明"为主题，是为公众提供科普教育的社会科技活动场所，科普旅游休闲的示范景点。

曜文
Glorious Culture

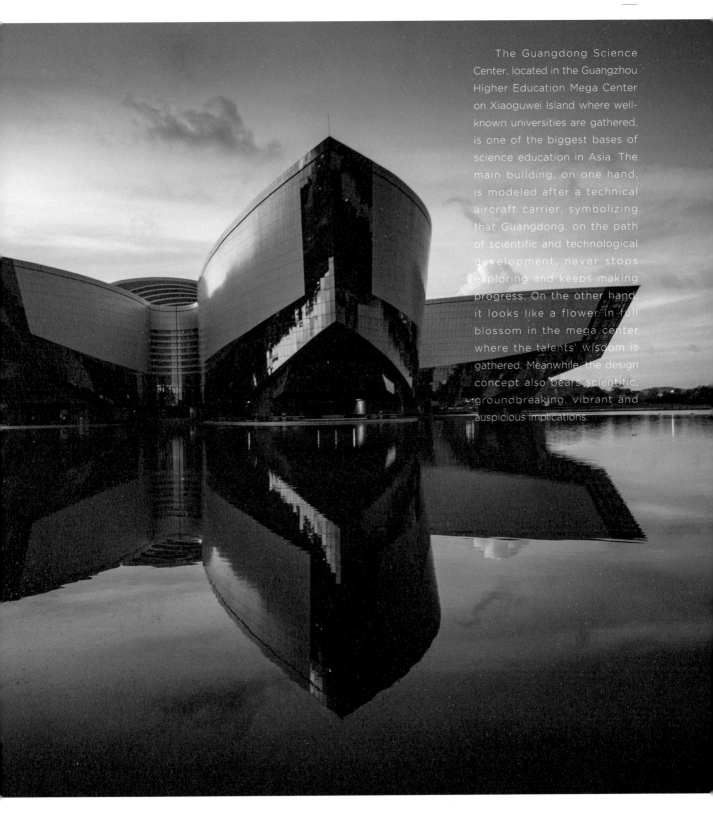

The Guangdong Science Center, located in the Guangzhou Higher Education Mega Center on Xiaoguwei Island where well-known universities are gathered, is one of the biggest bases of science education in Asia. The main building, on one hand, is modeled after a technical aircraft carrier, symbolizing that Guangdong, on the path of scientific and technological development, never stops exploring and keeps making progress. On the other hand, it looks like a flower in full blossom in the mega center where the talents' wisdom is gathered. Meanwhile, the design concept also bears scientific, groundbreaking, vibrant and auspicious implications.

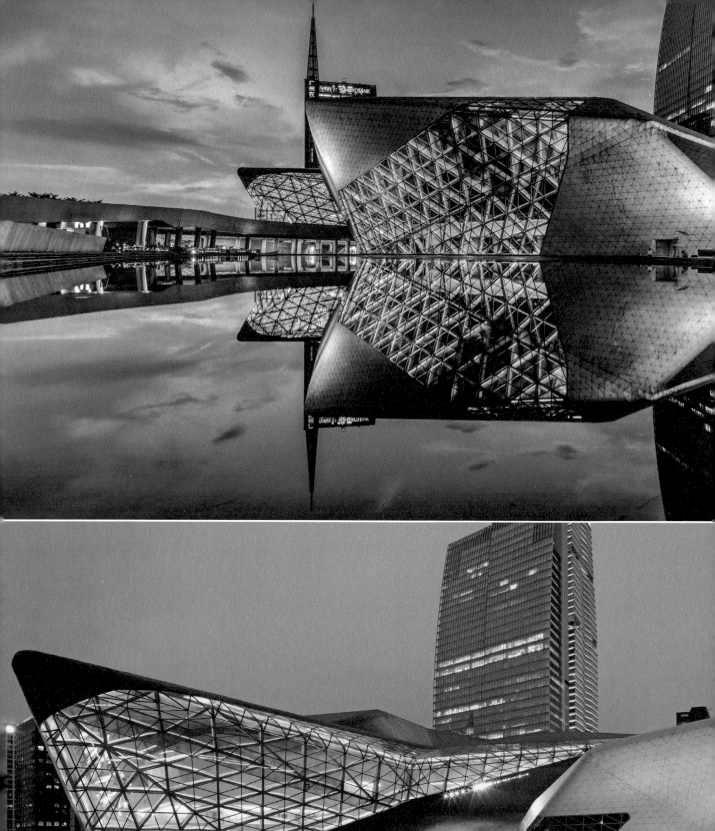

曜文
Glorious Culture

广州大剧院
Guangzhou Opera House

地址： 广州市天河区珠江西路1号
建成时间： 2010 年
建筑面积： 17.3万平方米
建筑规模： 歌剧厅共1840个座位，实验剧场共443个座位

Address: No.1 Zhujiang West Road, Tianhe District, Guangzhou
Completed in: 2010
Building area: 173,000 m²
Scale: The performance hall accommodates 1840 seats and the experimental theater accommodates 443 seats.

广州大剧院由世界著名的女建筑师扎哈·哈迪德主持设计，被外媒评为"世界十大歌剧院""世界最壮观剧院"，它有独特的建筑风格，充满未来感及流线美，其中尤以建筑与声学最高结合的"满天星"大剧场最为震撼，是广州新中轴线上当之无愧的标志性建筑之一，是目前华南地区最先进、最完善和最大的综合性表演艺术中心。

大剧院的造型宛如两块被珠江水冲刷过的灵石，被形象地称为"圆润双砾"。其中，"大砾石"是1800座的歌剧厅及其配套的设备用房、剧务用房、演出用房、行政用房、录音棚和艺术展厅；"小砾石"则是400座的多功能剧场及配套餐厅。建筑通过清晰完整的形体，连续匀质的界面（屋顶、外墙），没有时间尺度的材料（玻璃、抛光的金属）暗示了歌剧院与新城的密切联系。这里，建筑成为城市的艺术品，也强调了其公共性。

设计围绕公共艺术这一都市话题，提供的不是一般观赏性休闲绿地，而是一个生动的城市生活舞台，一个公众触摸艺术的界面。艺术广场设计力图延伸、叠合传统的建筑、广场、园林概念，使建筑不再是用来界定广场的边界，而成为广场及园景的延伸。

 Designed by world-famous architect Zaha Hadid, the Guangzhou Opera House is rated as one of the "10 best opera houses around the world" and "The world's most spectacular theaters" by foreign media. The Opera House is special in style and fluid in form. The performance hall has attracted much attention and respect for its lighting design which makes the ceiling like a starry sky, as well as for its perfect combination of the architectural design and the acoustic technologies.The Opera House, therefore, is one of the landmarks on the new central axis, and also the largest, the most advanced and well-equipped performing center in the southern China. Hadid, when designing the Opera House, was inspired by the story of the giant stone in the Pearl River smoothed by erosion. The two lopsided buildings are like a pair of boulders polished by the Pearl River; therefore, they are also known as "twin boulders". The larger boulder contains a 1,800-seat performance hall, and the smaller one houses a 400-seat multi-function theater. The irregular shape of the opera house makes it unique and creative, with every single dot, line and section bringing people endless imagination.

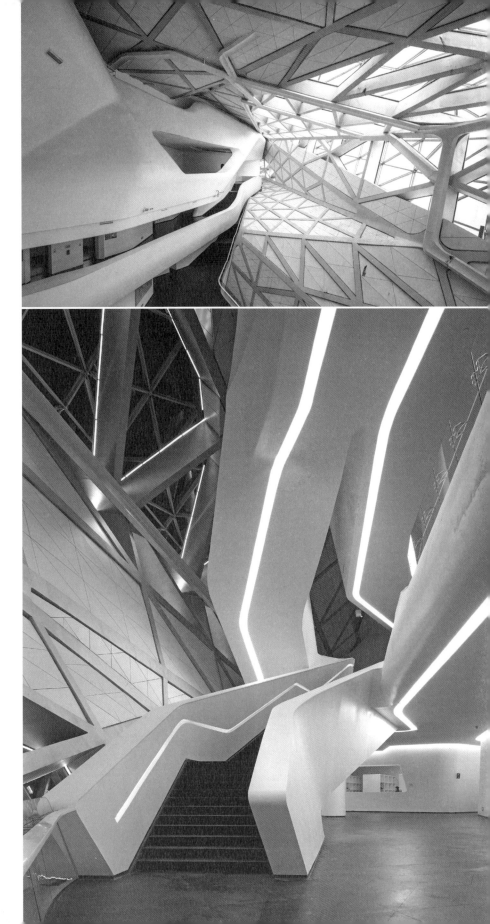

一域商都新韵
New Architecture

广东省博物馆
Guangdong Museum

地址：广州市天河区珠江东路2号	Address: No.2 Zhujiang East Road, Tianhe District, Guangzhou
建成时间：2010年	Completed in: 2010
占地面积：4.1万平方米	Floor area: 41,000 m^2
建筑面积：6.7万平方米	Building area: 67,000 m^2
层数：地下一层，地上五层	Height: 1 floor below ground, 5 floors above ground.

  广东省博物馆位于广州新城市中轴线中心区，集展览陈列、研究教育及文物保护等功能于一身，是国家一级博物馆，也是一座集建筑艺术和现代技术于一身的大型公共建筑。它的外形设计构思是将博物馆形象地比喻为一件摆放在绿丝绒上的宝盒，里面盛放着自然、文化和历史的宝物。而内部空间的设计理念则来源于广东传统的工艺品——象牙球，喻形于意，巧妙地将象牙球镂空的工艺运用在博物馆的空间组织上，使内部功能层层相扣，展示出多样的空间及通透感，带领着访客从内至外层层而进，功能流线自然而生。

  博物馆的巨型屋面钢桁架整体滑移总重量达8700吨，是世界重量最大的钢结构高空滑移工程。由于展馆是由巨型钢桁架悬吊出来的，中间没有一根结构柱，为陈列展示提供了很好的条件。

  广东省博物馆以稳重端庄、工整理性的形象构图，与广州歌剧院的灵动、圆润形成对比，为珠江新城中轴线上的文化艺术广场围合出亲切、完整的环境空间。

The Guangdong Museum, located at the center of the new central axis, is a national first-grade museum of China and a large public building revealing both the aesthetics of architecture and modern technology. The shape of the museum takes its inspiration from a treasure box on a green velvet carpet, which collects natural, cultural and historical treasures. The design of its interior space is derived from the traditional Cantonese ivory puzzle ball. The spatial arrangement of the museum is designed to follow the ivory-layered composition of traditional Cantonese crafted artwork, which allows each layer to be skillfully linked with one another. Thus, visitors can be directed to each floor by following the main circulation space, going from the atrium towards the periphery spaces.

南越王宫博物馆
Archaeological Site Museum of Nanyue Palace

地址：广东省广州市越秀区中山四路316号
建成时间：2010年
总建筑面积：16846平方米

Address: No.316, the forth Zhongshan Road, Yuexiu District, Guangzhou, Guangdong Province
Completed in: 2010
Total Building area: 16,846 m^2

广州南越王宫博物馆是一个遗址博物馆。秦朝末年，原秦将赵佗占据岭南三郡，于公元前203年建立南越国，定都番禺（广州），历五代，至公元前111年为汉所灭。经过近百年的经营，岭南地区得到很大的开发和发展。遗址自1995年开始发掘以来，先后发现了南越国时期的大型地下石构水池和南越国王宫御苑。其园林水景设计之独特，构筑之精巧，规模之宏大，令人赞叹不已，体现了先进的园林设计理念和造园要素，是目前国内保留最完好、时代最早的御苑遗迹。

值得一提的是，在南越王宫遗址的下层为秦造船遗址，在王宫遗址之上又有东汉至民国的12朝历史遗迹，它们犹如一部无字的史书，记载着广州2200多年的发展史。遗址位于广州市旧城中心最繁华的中山四路，说明广州的城市中心2200年来未曾移动过。这在中国历史上也是少见的。

南越王宫遗址先后两次被评为全国十大考古发现，对研究中国历史文化，研究中国古代城市（特别是古代广州）、古代建筑史和古代工艺史有极重要价值，是广州历史文化名城的精华所在。遗址已被国务院公布为全国重点文物保护单位。目前，考古发掘仍在进行中。

The Museum of Nanyue Palace is an archaeological site museum. At the end of Qin dynasty, a general of Qin dynasty, Zhao Tuo established the Nanyue Kingdom in 203 B.C. with Panyu(Guangzhou) as the capital. After about one hundred years, the Lingnan area achieved significant progress. Since the archaeological site was founded in 1995, we have discovered a large stone structural water pool and the royal palace of the Nanyue Kingdom. Its unique landscape design, precise structure and large scale are amazing, which have advanced landscape design concept and building elements, and they are the best preserved and the earliest archaeological sites of a royal palace. The archaeological work has been in progress.

曜文
Glorious Culture

广州新图书馆
New Guangzhou Library

地址：广州市天河区珠江东路4号
建成时间：2012 年
总建筑面积：9.8万平方米

Address: No.4 Zhujiang East Road, Tianhe District, Guangzhou
Completed in: 2012
Total Building area: 98,000 m^2

  广州新图书馆以"美丽书籍"为设计理念，寓意书籍的重叠和历史文化的沉积，同时融入骑楼等文化元素，体现了岭南建筑艺术特色。"我们的设计灵感来源于书籍，想象这个图书馆是由很多书籍堆积而成。同时考虑到旁边的博物馆是方形设计，我们作为紧密相连的建筑，最好以长条形并具线性美感的建筑匹配，因此才有大家看到的图书馆'之'字外形。"——参与广州新图书馆建筑设计的日本建筑师宫川浩这样介绍图书馆的设计。

  广州新图书馆是世界上规模最大的城市公共图书馆之一，藏书量达350万册。

一域商都新韵
New Architecture

Glorious Culture

On the new central axis of Zhujiang New Town, there stands a building that looks like an opened book. This is the Guangzhou Library, one of the largest urban public libraries in China, holding 3.5 million books. "The concept was inspired by books. We have pictured that this library is made up of a pile of books. At the same time, taking into account that the adjacent museum has a square-shaped design, and that the two structures will be in close proximity to one another, we decided that a proper design would be rectangular and linear in nature. With this in mind, the Chinese character-shaped (之) design was chosen", said Hiroshi Miyakawa, a Japanese architect who participated in the design of the Guangzhou Library.

悦宾
Hosp-
itality

广州是中国步入现代城市的先行者之一，在对外合作交流方面也是一马当先。山庄旅社、双溪别墅等，是20世纪60年代对园林酒店的探索，盛极一时，当时的国家领导人周恩来、陈毅等人都曾在山庄旅社接见外国首脑。1968年开业的广州宾馆，更是以27层的高度，成为当时全国最高的楼房。

改革开放初期，国家决定建设一批中外合作宾馆，国务院利用外资建设旅游饭店领导小组决定在广州建设3座五星级宾馆，于是，三大中外合资宾馆——白天鹅宾馆、中国大酒店、花园酒店相继开业，见证了广州对外开放的历史姿态和敢为人先的时代精神。迈入新世纪，更多的国际高端奢华酒店品牌入驻广州，W酒店、四季酒店、柏悦酒店等，都宣告广州已经发展成为四通八达的国际性贸易大都市。

近几年，随着广州酒店行业的迅猛发展，大量高档星级酒店建筑相继落成。为了提高市场竞争力，提升客人的体验感，广州最早的3家五星级酒店白天鹅宾馆、中国大酒店、花园酒店分别在不同时期做了彻底的升级改造，继续为广州酒店业创造新价值。

Guangzhou is one of the first modernized cities in China, and is also a leader in foreign cooperation and communication. Hotels including the Mountain Villa Guangzhou and the Shuangxi Villa demonstrate the efforts of building garden hotels in the 1960s. At that time, the state leaders including Zhou Enlai and Chen Yi met foreign heads of state in the Mountain Villa Guangzhou. The Guangzhou Hotel, opening in 1968, once ranked the top for its 27-story height.

At the beginning of the Reform and Opening-up, China decided to build a number of hotels eligible for accommodating foreign visitors. The Working Group on building hotels with foreign capital established by the State Council decided to build three five-star hotels in Guangzhou. As a result, the three major Sino-foreign joint venture hotels, namely the White Swan Hotel, the China Hotel and the Garden Hotel were set up one after another, which marked a testimony to Guangzhou's opening to the outside world and its pioneering spirit. Entering the new century, there are more international high-end hotel brands in Guangzhou, including the W Hotel, the Four Seasons Hotel and Park Hyatt Hotel, which declares that Guangzhou has become an international trade metropolis.

In recent years, with the rapid development of hotel industry of Guangzhou, a large number of star-rated hotels were built in Guangzhou. For improving market competitiveness and customers' experience, White Swan Hotel, China Hotel and Garden Hotel, the first three five-star hotels in Guangzhou, were upgraded and reconstructed during different periods. These old five-star hotels are continuing to create new value for the hotel industry in Guangzhou.

40 年来，广州的酒店业蓬勃发展，一座座优秀的建筑接连落成，向国际社会展示着中国南大门的面貌。

Over the past four decades, the hotel industry in Guangzhou has achieved vigorous development, with new buildings being completed successively, which has presented Guangzhou as the South Gate of China to the international community.

1978

- 1978 国务院利用外资建设旅游饭店领导小组决定在广州建设3座五星级宾馆
- 1978 改革开放城市重心逐渐东移
- 1981 全国第一个超级商场——广州友谊公司超级自选商场开业
- 1982 广州入选第一批国家级历史文化名城
- 1983—1985 三大中外合资宾馆先后建成开业
- 1984 广州经济技术开发区成立
- 1984 广州图书馆开馆

- 1983 白天鹅宾馆 White Swan Hotel
- 1984 中国大酒店 China Hotel
- 1984 花园酒店 The Garden Hotel

1988

- 1987 第六届全国运动会在广州召开
- 1987 广东奥林匹克体育馆动工兴建
- 1985 天河体育中心建成
- 1990 地铁首期工程获批
- 1993 开始建设珠江新城
- 1994 正佳广场奠基
- 1996 作为内地最早一批商城的天河城开业
- 1997 广州地铁开通第一条线路
- 1998 广东奥林匹克体育馆动工兴建
- 1997 广州公路主枢纽首个大型公用客运站——天河客运站建成
- 1999 天河城广场全面开业

- 1987 天河体育中心 Tianhe Sports Center
- 1988 西汉南越王墓博物馆 Museum of the Mausoleum of the Nanyue King
- 1996 天河城 Teemall
- 1997 中信广场 CITIC Plaza
- 1997 广东美术馆 Guangdong Art Gallery

1998

- 2000 启动《广州城市建设总体战略概念规划纲要》，提出"南拓、北优、东进、西联"八字方针
- 2001 《广州城市建设总体战略概念规划纲要》通过
- 2001 第九届全国运动会在广州召开
- 2002 通过并正式实施南沙开发区规划
- 2004 广州成功获得2010年亚运会的举办权
- 2004 珠江新城轴线四大公建计划全面启动
- 2005 国务院批复《广州市城市总体规划（2001—2010年）》
- 2005 正佳广场、太古汇、万菱广场等一系列大型商城相继建成开业
- 2005 广州地铁3号线开通
- 2006《总规》修编，在八字方针基础上增加"中调"战略
- 2007《广州2020:城市总体发展战略规划》开展

- 2001 广东奥体中心 Guangdong Olympic Sports Center
- 2001 广州新体育馆 Guangzhou Gymnasium
- 2002 琶洲国际会展中心 Guangzhou International Convention and Exhibition Center
- 2003 逸夫人文馆 Shaw Building of Humanities
- 2004 广州白云国际机场 Guangzhou Baiyun International Airport
- 2005 正佳广场 Grandview Mall
- 2006 保利国际广场 Poly International Plaza
- 2007 白云国际会议中心 Guangzhou Baiyun International Convention Center

2008

- 2008 广东科学中心 Guangdong Science Center
- 2010 广州国际金融中心(西塔) Guangzhou International Finance Center (West Tower)
- 2010 广州塔 Canton Tower

20世纪70年代的白鹅潭边
The Baietan in the 1970s

1983年春节，白天鹅宾馆正式营业并对社会开放
In the Spring Festival of 1983, the White Swan Hotel was officially opened to the public

花园酒店的原址，图中大楼为白云宾馆
The former location of Garden Hotel; the building in the picture is Baiyun Hotel

20世纪80年代初花园酒店建设工地
The construction site of Garden Hotel in the early 1980s

2017 《深化粤港澳合作 推进大湾区建设框架协议》签署
2017 广州《财富》论坛举办
2016 作为天河路商圈收官之作的天环广场正式开业
2015 提出打造三大枢纽、完成一江两岸三带的目标
2014 《广州市历史文化名城保护规划》获批复
2011 国务院正式批复《广州市土地利用总规划（2006—2020年）》
2010 首届亚洲残疾人运动会在广州开幕
2010 广州亚运城竣工
2010 第16届亚洲运动会在广州开幕灯
2010 广州塔亮灯
2009 粤剧列入世界非物质文化遗产名录

2018
2018 广州城市规划展览馆 Guangzhou Urban Planning Exhibition Center
2017 保利天幕广场（琶洲眼）Poly Skyline Square (Pazhou Eye)
2016 周大福金融中心（东塔）Chow Tai Fook Financial Center (East Tower)
2015 天环广场 Parc Central
2015 越秀金融大厦 Yuexiu Financial Tower
2013 珠江城（烟草大厦）Pearl River Tower (Tobacco Building)
2012 W酒店 W Hotel
2012 广州新图书馆 New Guangzhou Library
2012 富力盈凯广场 R&F Yingkai Square
2012 利通大厦 Leatop Plaza
2011 广州太古汇 Taikoo Hui
2011 广州国际大厦 The Pinnacle (Guangzhou)
2010 广州南站 Guangzhou South Railway Station
2010 南越王宫博物馆 Archaeological Site Museum of Nanyue Palace
2010 广东省博物馆 Guangdong Museum
2010 广州大剧院 Guangzhou Opera House
2010 南沙体育馆 Nansha Gymnasium
2010 广州国际体育演艺中心 Guangzhou International Sports Arena
2010 广州亚运城综合体育馆 Asian Games Town Gymnasium
2010 广东奥体中心游泳跳水馆 Guangdong Olympic Sports Center Swimming and Diving Hall
2010 海心沙亚运公园 Haixinsha Asian Games Theme Park
2010 广州圣丰广场 Guang-zhou Shengfeng Plaza

悦宾
Hospitality

白天鹅宾馆
White Swan Hotel

地址： 广州市沙面南街1号
建成时间： 1983年
总建筑面积： 10万平方米
客房数： 1000间

Address: No.1 Shamian South Street, Guangzhou
Completed in: 1983
Total Building area: 100,000 m²
Rooms: 1000

1979年，白天鹅宾馆的诞生有其独特的历史背景，它是改革开放的一个坐标，具有里程碑的意义。1983年建成开业的白天鹅宾馆有着多个"第一"的头衔：我国对外开放后第一家粤港合作经营的宾馆，全国第一家由中国人自行设计、施工、管理的大型现代化酒店，内地第一家五星级酒店，并被世界一流酒店组织接纳为内地的首家成员。

白天鹅宾馆由岭南著名建筑师佘畯南、莫伯治主持设计。它的建筑外观表现出浓郁的现代化气息，主楼外观平面呈扇状腰鼓形，有天鹅羽翼重叠之意。而内部空间则全方位地展现了中国文化特有的魅力，尤其是"故乡水"的设计和立体化共享空间的庭园打造受到了业内外人士的好评。

白天鹅宾馆利用现代技术实现了现代主义建筑与传统岭南庭园空间的有机结合，取得了极大的声誉，被认为是中国现代主义建筑的"新景象"，是现代岭南建筑的代表作之一。

2012—2015年的升级改造，让白天鹅宾馆再次成为广州最高端商务酒馆之一。

Completed and opened in 1983, the White Swan Hotel has several titles being rated as "the first": the first hotel jointly operated by Guangdong Province and Hong Kong businesses after China's Reform and Opening-up, the first modern hotel designed, constructed and managed by China, the first five-star hotel in Chinese Mainland, and the first member of "The Leading Hotels of the World" in Chinese Mainland. It was designed by two renowned Lingnan architects, She Junnan and Mo Bozhi.

The exterior of the Hotel exudes a strong sense of modernism, while the interior space is full of Chinese cultural flavor, especially the design of the open courtyard and the "Hometown Water", an indoor cascading waterfall garden, which have been well received by both professionals and amateurs. It is considered as the new image of Chinese modern architecture and one of the representative works of modern Lingnan architecture.

一域商都新韵
New Architecture

中国大酒店
China Hotel

地址：广州市流花路122号
建成时间：1984年
建筑面积：15.9万平方米
客房数：888间

Address: No.122 Liuhua Road, Guangzhou
Completed in: 1984
Building area: 159,000 m^2
Rooms: 888

  中国大酒店是一座引进外资兴建的具有世界一流水平的现代化酒店，是当时中国首批、华南地区唯一的"白金五星级"酒店。它雄踞越秀山麓，尽占地理优势，是一个集高级旅馆、高级公寓、商业大厦及相应附属建筑于一身的综合性多功能酒店，当时在国内尚属首创。

  整个建筑群外观宏伟、整齐，低层部分互相沟通快捷便利。设计上采用了高低相间的空间组合形式，在低层屋面遍种花木，设置中式庭院。庭院中,小桥流水，石山飞瀑，曲径通幽，有强烈的地域特色，在高密度的建筑群中，开辟了一片自然的天地。

  中国大酒店还有一大特色是西端入口两面山墙上的大型线雕壁画（创作者：潘鹤先生），两幅壁画的面积达千余平方米，采用线描的形式在混凝土上面直接刻线，然后在凹线上贴金箔。两幅壁画题材分别是"歌舞庆升"和"贸易通四海"，再现了广州与海外交流的传统，透出广州由来已久的"商文化"气息。

Located on Liuhua Road in the central Yuexiu District of Guangzhou, the China Hotel is a multi-function hotel consisting of high-end hotel rooms and apartment, commercial buildings and other ancillary buildings. It is the first modern five-star hotel that has been built by utilizing foreign capital. The hotel is generally well-arranged with a strong national style. The design embodies traditional style of Chinese architecture and adopts an alternating arrangement of high and low buildings. Flowers and trees are planted all over the low-rise roof. Courtyards of Chinese style and modern architectural techniques are skillfully integrated. The building is solemn in shape and diverse in style. It reflects both the features of Chinese architecture and the Zeitgeist.

一域商都新韵
New Architecture

花园酒店
The Garden Hotel

地址：广州市越秀区环市东路368号
建成时间：1984 年
总建筑面积：17万平方米
客房数：2140间

Address: No.368 Huanshi East Road, Yuexiu District, Guangzhou
Completed in: 1984
Total Building area: 170,000 m^2
Rooms: 2140

　　花园酒店是20世纪80年代初广州的重要建筑物。它的"Y"字形平面、顶层旋转餐厅及总统套房等做法在国内建筑界曾引起轰动效应。

　　花园酒店的外观朴素无华，没有多余的装饰，但两个塔楼立面具有比例恰当的矩阵式布窗格，与其他6个实墙立面形成了对比。两个"Y"字形平面前后错落创造出丰富的空间构成。入口处跨度约40米的壳体镂空大雨棚更是别出心裁，构成了花园酒店前庭的视觉中心。

　　酒店的大堂设计融合中西文化艺术特色，拥有全国最大的"红楼梦"大理石贴金壁画、"广东水乡风貌"大理石壁画以及其他漆画。大堂正中央天花板镶有广东省最大的"金龙戏珠"藻井，仿照中国古代皇宫殿顶设计。酒店的后花园以"天涯若比邻"为主题，面积达1万平方米，有大型壁画、假山瀑布、亭台水榭及草木，别具岭南园林风格特色。

悦宾
Hospitality

The Garden Hotel is a "Platinum Five-Star Hotel" which has 2,140 rooms and provides services including accommodation, dining, business, conference and tourism. The design of the Y-shaped building, the revolving restaurant and the presidential suites have had great influences on China's architecture community. It made the hotel an important construction in Guangzhou in the early 1980s. The design of the lobby reflects both Chinese and Western cultural and artistic elements. Large murals, indoor rockery and waterfall, pavilions and plants have formed a typical style of Lingnan garden.

悦宾
Hospitality

The Guangzhou Shengfeng Plaza stands in the Tianhe North CBD. The Phase I consists of super-grade office building and the Sofitel five-star hotel; and the Phase II includes international business apartment and a waterfront-themed CBD bar street. The plaza, designed by SBA, a U.S. Design Firm, is magnificent in its general look and sharp in details. It pioneers high-quality commercial buildings in Guangzhou with its forward-looking design concept, dynamic streamline shape, 24-hour global office support services, world-class configuration and 8 advanced functions.

广州圣丰广场
Guangzhou Shengfeng Plaza

地址：广州市天河区广州大道中988号
建成时间：2010年
总建筑面积：15.5万平方米

Address: No.988 Guangzhou Avenue, Tianhe District, Guangzhou
Completed in: 2010
Total building area: 155,000 m²

圣丰广场矗立于天河北CBD核心商务圈，一期由超甲级办公楼和五星级酒店组成；二期由国际商务公寓和水岸文化主题CBD酒吧街区组成，由美国SBA设计集团担纲设计，大气中不失精致，细节中显露锋芒。

针对用地紧张的场地特征，规划上采用流线型的平面及立面处理，办公楼以椭圆形平面为蓝本逐层收缩，形成高耸挺拔的形象。酒店以"L"形平面和办公楼组合，入口采用一个大跨度薄壳玻璃雨棚，适应地方气候特点，形成了良好的外部空间，塑造出雄浑大气的整体形象，成为天河商务中心的一道亮丽风景线。

113

悦宾
Hospitality

W 酒店
W Hotel

地址：广州市天河区兴盛路7号
建成时间：2012年
总建筑面积：10.65万平方米

Address: No.7 Xingsheng Road, Tianhe District, Guangzhou
Completed in: 2012
Building area: 106500 m^2

广州W酒店作为中国内地首家W酒店，地处珠江新城繁华地段，包含一个精品酒店和一所服务式公寓为主的综合体。建筑设计既是对广州珠江新城城市形态的回应，亦是紧凑式都市酒店功能的独特演绎。

W酒店精妙融汇各种建筑设计元素，以新颖时尚的方式描绘出广州这座古老城市在现代化工业与经济爆发式增长过程中所处的独特地位，它的整体建筑融入城市肌理，造型大胆地采用不对称外观设计。

W酒店时尚的黑色玻璃外衣下穿插着精心安排的镂空和明亮玻璃元素的点缀，在立面和入口的处理上刻意强调了酒店私人领域和公共领域的对比，将酒店气氛向外散播，并以垂直玻璃条分隔每个房间，营造亲切的尺度。色彩、灯光与一层又一层的结构，把W酒店打造成一曲充满活力、时尚别致的现代交响乐。

Located in the bustling area of the Zhujiang New Town, the W Guangzhou is the first W hotel in Chinese mainland. It is an amalgamation of a boutique hotel and serviced apartments. The design echoes the urban landscape of Zhujiang New Town, and also redefines the function of a compact hotel in urban area. The W Guangzhou has unified various architectural elements in its design: a striking exterior with an asymmetrical shape, a fashionable cover in black glass punctuated by architectural cut-outs and bright glass highlights, multiple colors, lighting and layers of structure. All of them have translated the hotel into a vibrant and special modern symphony.

繁津

Downtown Development

广州号称"千年商都",它是海上丝绸之路的重要起点。早在唐宋,广州就是我国第一批由商业发展起来的商业城市,丝绸、陶瓷和茶叶就是由此走向世界。到了宋元明时期,广州港更成为国际著名的贸易港口,号称"东方第一大港"。即便到了清朝闭关锁国时期也保留了广州通商,当时的十三行富可敌国,将广州的商业繁荣推上了顶峰。

而在新中国成立之时,受到国际封锁,广州成为我国唯一进出口城市,广交会1957年创办至今从未中断,几乎浓缩了一部中国外贸发展史,可以说商贸是广州的根和魂。改革开放伊始,珠三角走在最前线,广州更是遇到了发展的黄金时期,80年代西湖灯光夜市、高第街第一批商场在羊城诞生,宣告着市场经济的到来。随着城市发展战略重心的转移,广州商业版图不断扩大。千年不衰,广州成了商都中的传奇。

20世纪80年代以来,城市的重心逐渐东移到天河,广州建设起一座座大型商业综合体。1996年开业的天河城是当时中国最早购物中心之一,开启了一个全新购物时代。1997年广州正式开通了第一条地铁线路,人们出行的便利催生了更多大型购物商场在地铁附近开业。 此后正佳广场、广州太古汇、天环广场、高德置地广场和 K11 购物艺术中心相继落成,地铁3号线的建成将它们联系成了融购物、休闲、娱乐为一体的天河商圈。

广州的城市格局发展,与经济社会建设相辅相成,共同推进。2004年广交会落户琶洲国际会展中心,带动琶洲的商贸发展,保利国际广场、保利天幕广场等一批出色的商业综合体为广州增色不少,成为广州新时代的标志之一。

Known as a city of commerce for a thousand years, Guangzhou is an important starting point of the Maritime Silk Road. As early as in the Tang (618-907) and Song (960-1279) Dynasties, Guangzhou was one of the earliest cities flourished by its commercial development. Silk, china and tea were therefore introduced to the world. In the Song, Yuan and Ming Dynasties (960-1644), the Port of Guangzhou became a world-famous trade port, known as the busiest port in the East. Even during the period of the isolationist foreign policy in the Qing Dynasty (1644-1911), Guangzhou was made the port open to foreign commerce. The Thirteen Factories of that time were incredibly rich. Guangzhou reached a peak in terms of its commercial prosperity.

When the People's Republic of China was founded, China suffered an international embargo. Guangzhou was made the only city in China for import and export. The China Import and Export Fair (or Canton Fair) began in 1957 and is now still held every year, which witnessed the history of China's foreign trade development. It can be said business and trade are the root and soul of Guangzhou. At the beginning of the Reform and Opening-up, the Pearl River Delta took a lead in opening up to the outside world, and Guangzhou also entered a golden era for its development. In the 1980s, the first batch of markets including the Xihu Road Lamplight Night Market and the Gaodi Street were born in Guangzhou, which announced the beginning of the market-oriented economy. With the focus shift of the city's development strategy, Guangzhou's commercial development has been expanded. After more than a thousand years of development, Guangzhou has become a legendary city of commerce.

Since the 1980s, the city center has gradually moved eastward to Tianhe District; and a number of large-scale commercial complexes have been built in the city. The Teemall, opening in 1996, was one of the earliest shopping centers in China of the time, which unfolded a new era of shopping centers. In 1997, Guangzhou Metro Line 1 officially became operational. The convenience of traveling has enabled more large-scale shopping malls to be built near the metro station. Since then, the construction of shopping centers including the Grandview Mall, the Taikoo Hui, the Parc Central, GT Land Plaza and Guangzhou K11 Art Mall have been completed one after another. They are then all linked by the Metro Line 3 and form the Tianhe Commercial Circle which is an amalgamation of shopping, leisure and entertainment.

The urban development and the economic and social construction in Guangzhou are mutually reinforced. In 2004, the Canton Fair was held in the Guangzhou International Convention and Exhibition Center, which has driven the commercial development of the Pazhou Island. In addition, a series of business complexes, including the Poly International Plaza and the Poly Skyline Square have made contributions to the city's development and become landmarks of Guangzhou in the new era.

广州，自古有商贸都市的美誉。社会主义市场经济确立以来，广州涌现出一批又一批的批发市场、商业综合体。时代的印记深深刻在广州的商贸都市版图上，见证着中国经济的崛起和腾飞。

Guangzhou has traditionally been the center of business and trade. Since the establishment of the socialist market economy, a number of wholesale markets and business complexes have sprung up. Guangzhou, a commercial and trade metropolis, has been engraved with the mark of the times and has witnessed the rise and take-off of China's economy.

1978

- 1978 国务院利用外资建设旅游饭店领导小组决定在广州建设三座五星级宾馆
- 1978 改革开放城市重心逐渐东移
- 1981 全国第一个超级商场——广州友谊公司超级自选商场
- 1982 广州入选第一批国家级历史文化名城
- 1982 广州图书馆开馆
- 1983–1985 三大中外合资宾馆先后建成开业
- 1984 广州经济技术开发区成立
- 1983 白天鹅宾馆 White Swan Hotel
- 1984 中国大酒店 China Hotel
- 1984 花园酒店 The Garden Hotel

1988

- 1987 广东奥林匹克体育馆动工兴建
- 1987 第六届全国运动会在广州召开
- 1985 天河体育中心先行建成开业
- 1987 广州地铁开通第一条线路
- 1996 作为内地最早一批商城的天河城开业
- 1994 正佳广场奠基
- 1993 开始建设珠江新城
- 1990 地铁首期工程获批，全长12.7公里
- 1998 广州公路主枢纽首个大型公用客运站——天河客运站建成
- 1997 天河体育中心 Tianhe Sports Center
- 1987 南越王墓博物馆 Museum of the Mausoleum of the Nanyue King
- 1988 西汉南越王墓博物馆
- 1996 天河城 Teemall
- 1997 中信广场 CITIC Plaza
- 1997 广东美术馆 Guangdong Art Gallery

1998

- 2000 启动《广州城市建设总体战略概念规划纲要》，提出"南拓、北优、东进、西联"八字方针
- 2001 《广州城市建设总体战略概念规划纲要》通过
- 2001 第九届全国运动会在广州召开
- 1999 提出珠江新城轴线规划
- 1999 天河城广场全面开业
- 2002 通过并正式实施南沙开发区规划
- 2004 广州成功获得2010年亚运会举办权
- 2004 珠江新城轴线四大公建计划全面启动
- 2005 广交会正式移师琶洲会展中心
- 2005 国务院批复《广州市城市总体规划（2001—2010年）》
- 2005 正佳广场、太古汇、万菱汇等一系列大型商城相继建成开业
- 2006 《总规》修编，在八字方针基础上增加"中调"战略
- 2007 《广州2020：城市发展战略规划》开展
- 2005 广州地铁3号线开通
- 2001 广州奥体中心 Guangzhou Olympic Sports Center
- 2001 广州新体育馆 Guangzhou Gymnasium
- 2002 广州国际会展中心 Guangzhou International Convention and Exhibition Center
- 2003 逸夫人文馆 Shaw Building of Humanities
- 2004 广州白云国际机场 Guangzhou Baiyun International Airport
- 2005 中佳广场 Grandview Mall
- 2006 保利国际广场 Poly International Plaza

2008

- 2008 广东科学中心 Guangdong Science Center
- 2007 白云国际会议中心 Guangzhou Baiyun International Convention Center
- 2010 广州国际金融中心（西塔） Guangzhou International Finance Center (West Tower)
- 2010 广州塔 Canton Tower

创办于1914年的先施公司是广州第一间百货公司
The Sincere Company Ltd., the first department store in Guangzhou founded in 1914

20世纪90年代广州最繁华的商业区南方大厦
The Nanfang Building in the busiest commercial district of Guangzhou in the 1990s

1996年天河城试业典礼
The trial operation ceremony of the Teemall in 1996

2005年正佳广场开业
The Grandview Mall opened in 2005

建设中的天环广场
The Parc Central in construction

天河商圈
The Tianhe Commercial Circle

2017 《深化粤港澳合作 推进大湾区建设框架协议》签署
2017 广州《财富》论坛举办
2016 作为天河路商圈收官之作的天环广场正式开业
2015 提出打造三大枢纽、完成一江两岸三带的目标
2014 《广州市历史文化名城保护规划》获批复
2011 国务院正式批复《广州市土地利用总体规划（2006—2020年）》
2010 首届亚洲残疾人运动会在广州开幕
2010 第16届亚洲运动会在广州开幕
2010 广州塔亮灯
2010 亚运城竣工
2009 粤剧列入世界非物质文化遗产名录

2018

2018 广州城市规划展览馆
Guangzhou Urban Planning Exhibition Center
2017 保利天幕广场（琶洲眼）
Poly Skyline Square (Pazhou Eye)
2016 周大福金融中心（东塔）
Chow Tai Fook Financial Center (East Tower)
2015 天环广场
Parc Central
2015 越秀金融大厦
Yuexiu Financial Tower
2013 珠江城（烟草大厦）
Pearl River Tower (Tobacco Building)
2012 W酒店
W Hotel
2012 广州新图书馆
New Guangzhou Library
2012 富力盈凯广场
R&F Yingkai Square
2012 利通大厦
Leatop Plaza
2011 广晟国际大厦
The Pinnacle (Guangzhou)
2011 天环太古汇
Taikoo Hui
2010 广州南站
Guangzhou South Railway Station
2010 南越王宫博物馆
Archaeological Site Museum of Nanyue Palace
2010 广东省博物馆
Guangdong Museum
2010 广州大剧院
Guangzhou Opera House
2010 南沙体育馆
Nansha Gymnasium
2010 广州国际体育演艺中心
Guangzhou International Sports Arena
2010 广州亚运城综合体育馆
Asian Games Town Gymnasium
2010 广东奥体中心游泳跳水馆
Guangdong Olympic Sports Center Swimming and Diving Hall
2010 海心沙亚运公园
Haixinsha Asian Games Theme Park
2010 广州圣丰广场
Guangzhou Shengfeng Plaza

天河城
Teemall

地址：广州市天河路208号
建成时间：1996 年
总建筑面积：311115平方米

Address: No.208 Tianhe Road, Guangzhou
Completed in: 1996
Total building area: 311,115 m^2

　　天河城的建设历时 20 年，是我国第一个大型城市商业建筑综合体。它是我国第一个真正意义的shoppingmall，培养了我国第一代商业地产的开发管理人才，在行业中至今享有盛誉。它见证了广州的商业繁荣，见证了中国内地进入一个全新的消费时代，被誉为"中国第一商城"。

　　1996年天河城购物中心诞生之初，天河片区只是一片名为"苗圃"的荒芜之地，如今，曾经的"苗圃"已崛起成为天河路商圈，而作为开荒者的天河城，20年来仍风光无限，日平均客流量达30万人次，营业额一直名列前茅，缔造了世界零售业的财富神话。天河城开创了一种全新的消费理念，它的成功带来了"天河城效应"，带动了周边地带的繁荣，把广州的商业提高到一个新水平，成为后来的"造mall者"必定朝拜的"圣地"。

　　天河城的设计力图通过简洁有力的造型和适度的比例控制，表现时代感、塑造标志性。2008年和2012年，天河城办公楼和酒店也相继开业，形成大型的城市综合体。

The Teemall, which was built 20 years ago, is the first large-scale commercial complex and the earliest shopping mall in China that trained the first generation of elites in commercial real estate development and management Enjoying a high reputation in the industry, Teemall, known as the top mall in China, has witnessed the business boom of Guangzhou and the Mainland entering a new era of consumption. Teemall, by its successful business practices, has created a new idea of consumption and the "Teemall Effect". It has thereafter driven the growth of its neighboring area, elevating Guangzhou's business development to a new level.

繁津
Downtown Development

正佳广场
Grandview Mall

地址：广州市天河路228号　　　　Address: No.228 Tianhe Road, Guangzhou
建成时间：2005年　　　　　　　　Completed in: 2005
建筑面积：42万平方米　　　　　　Building area: 420,000 m²

　　正佳广场坐落于繁华的天河路商圈，以体验式主题购物乐园为设计定位，是当今中国最大的完全贯彻"体验式消费"模式，集零售、休闲、娱乐、餐饮、会展、康体、旅游及商务于一身的大型现代化购物中心。

　　正佳广场由美国著名的捷得事务所设计，它采用剧场式的商业空间布局，主中庭散发出一条螺旋形购物流线，"X"形串联了8个大小、形状各异的小中庭，分别连接四大出入口，形成集中与分散结合的"回"字形商业动线。

　　正佳广场领先全球，引入了前所未有的先进体验经济模式。商场划分为22个独特主题展区，其中包括全球首创的360°海底隧道、广州首家商场内的自然科学博物馆，以及引领潮流的极地海洋世界等。正佳广场突破传统购物概念，将购物与"体验"相融合，使其成为与拉斯维加斯、纽约第五大道、东京银座等世界一流购物区域媲美的体验经济巨人。

Situated in the busy Tianhe Road Commercial Circle, the Grandview Mall is designed and positioned as a shopping theme park with its focus on experience. Now, it is the largest modern shopping center that adopts a model of "experience consumption" and integrates services including retailing, leisure, entertainment, dining, exhibition, health center, tourism and business. By integrating shopping with experience, the Grandview Mall has become an experience economy giant, enjoying equal popularity with Las Vegas, the 5th Avenue in New York and the Ginza in Tokyo. It is the world's fifth largest and Asia's largest emporium.

Downtown Development

保利国际广场
Poly International Plaza

地址：广州市海珠区琶洲阅江中路688号 建成时间：2006 年 占地面积：5.7万 平方米 建筑面积：19.6万 平方米	Address: No.688 Yuejiang Middle Road, Pazhou Island, Haizhu District, Guangzhou Completed in: 2006 Floor area: 57,000 m^2 Building area: 196,000 m^2

  保利国际广场位于琶洲经济圈核心位置，北望珠江，被琶洲塔公园、亲水公园和体育公园三面环绕，具有良好的景观，是琶洲国际会展中心周边唯一的高端商务办公商业项目。保利地产集团聘请了美国SOM建筑事务所做国际广场建筑设计，美国SWA公司做园林景观设计，美国DPI公司做灯光设计，力求以建筑与景观最佳结合来达到城市名片的效果。

  "超长板式设计"使保利国际广场塔楼高宽比达到1：8，为目前国内之最，中间无一梁一柱，采用全角度采光设计，形成毫无遮挡的360°景观空间，开目前国内办公楼的先河。建筑围合而成的两万多平方米的大型水景中央园林，充分利用华南植被与气候，用简明的现代风格与传统中国园林，巧妙运用借景，创造出一个多层次、多尺度的怡人空间。

The Poly International Plaza, a key project of Poly Real Estate Group, was put into full use at the end of 2006. Located at the core of Pazhou central exhibition economic circle, this project has an advantageous landscape and is the only project for high-end commercial office use around the Pazhou International Convention Exhibition Center. The Poly International Plaza was designed by SOM, a U.S. architecture firm. It strives to become an icon of the city by combining the architecture with the landscape. In fact, the completion of the project also underscores the potential and artistic value of this international-level design.

广州太古汇
Taikoo Hui

地址：广州市天河路383号 建成时间：2011年 总建筑面积：45万平方米	Address: No.383 Tianhe Road, Guangzhou Completed in: 2011 Total building area: 450,000 m²

太古汇位于广州市天河中央商务区核心地段，由一个大型购物商场、两座甲级办公楼、一个文化中心、文华东方酒店及酒店式服务住宅构成。该项目由世界知名的建筑设计事务所Arquitectonica设计，定位为集休闲娱乐、商贸活动、文化艺术于一身的综合商业体。

椭圆的中庭、横贯商场顶层的玻璃天廊是广州太古汇商场独有的特色。在对光线的运用上，太古汇别具匠心，多个椭圆形镂空的"天井"使自然光直接洒落在各楼层，光照让地面以下的楼层不再具有压抑感。室内灯光以鹅黄色为主色调，而中庭底部连贯灯槽与手扶电梯下方如繁星般排列的小灯，则映衬着主色调进行同色系灯光点缀，使商场整体照明在白天自然舒适，夜间柔和雅致。

2017年，广州太古汇获得了美国绿色建筑委员会颁发的能源与环境设计先锋评级（LEED）当中"既有建筑运营与保养（EBOM）"类别的铂金级认证，成为全球第一个获得这个级别认证的室内商场。

一域商都新韵
New Architecture

Downtown Development

Taikoo Hui, officially opening in Guangzhou in 2011, is the largest investment project of Swire Properties in China. It was designed by Arquitectionica, a world acclaimed architectural design company, and developed and managed by Swire Properties. It is a large-scale multi-faceted complex composed of super high-rise office buildings, the Mandarin Oriental Hotel, Guangzhou Cultural Center and commercial podium buildings.

Drawing on Swire Properties' years of experience in developing international commercial buildings, the Taikoo Hui has become a new icon in Guangzhou for leisure, entertainment, commerce and cultural and artistic appreciation.

繁华
Downtown Development

天环广场
Parc Central

地址：广州市天河路218号	Address: No.218 Tianhe Road, Guangzhou
建成时间：2015年	Completed in: 2015
建筑面积：11万平方米	Total building area: 110,000 m²

天环广场坐落于广州天河中央商务区核心地段，傲踞广州中轴线焦点位置，融购物休闲、交通枢纽及城市空间为一体的城市商业综合体，首次将"公园"融入现代购物商场中的崭新设计，完美诠释人、建筑、环境与自然和谐相处的可持续发展理念，成为广州最具代表性的商业新地标。

从空中俯瞰天环广场整体建筑犹如跃动的双鱼造型，其设计理念来自中国文化中象征和平、吉祥和财富的鲤鱼，颇有"太极双鲤，鱼跃天河"的韵味，寓意城市的和谐、均衡和繁荣。有别于其他"盒子"形的商业项目，天环广场打破"围墙"的设计定式，地面保留大面积户外空间，配合层叠起伏的多层次绿化景观，如城市中的绿洲，为市民和游客创造出一片休闲社交与购物娱乐的活力时尚天地。

Located in the core area of the new CBD in Guangzhou. The Parc Central, an open shopping park consisting of buildings with levels both above and below ground, presents a new model for the fast-growing cities in the Pearl River Delta. This retail shopping mall is built along the main avenue, which has added a new visual element to the city. It is a complex leading the trend of retailing, transportation development and design of public building. The architecture, symbolizing the auspicious 'Double Fish' of traditional Chinese culture, represents harmony, peace, and fortune as it appears to leap over the Tianhe District like two fish.

保利天幕广场（琶洲眼）
Poly Skyline Square (Pazhou Eye)

地址：广州市海珠区琶洲阅江东路
建成时间：2017 年
占地面积：3.4万平方米
总建筑面积：31万平方米

Address: Yuejiang East Road, Pazhou Island, Haizhu District, Guangzhou
Completed in: 2017
Floor area: 34,000 m^2
Total building area: 310,000 m^2

  保利天幕广场位于广州琶洲岛中部珠江南岸，琶洲经济圈的核心位置。它的定位是地标型商业综合体项目，主打金融行业，尤其注重世界500强、中国500强企业客户的引进，同时也着重考虑引进新兴金融、互联网科技企业。SOM的设计师介绍说：设计的出发点，是尽可能将城市景观、最好的视野融入产品中。因此，被人们形象称为"琶洲之眼"。它拥有311米的高度，包括一栋38层的酒店、一栋60层的办公楼，和6层的酒店裙房。双塔楼独特的几何形体及顶部大气的开口，呈朝向珠江新城和珠江的双"弓"形，成为一览众山小的制高点，为广州这座国际大都市的天际线增色不少，成为广州城的标志性建筑。

  The Poly Skyline Square is located in the heart of Pazhou Island on the south bank of the Pearl River, the core area of Pazhou economic circle. It is 311 meters in height and includes a 38-floor hotel, a 60-floor office building and a 6-floor podium building. The architect from SOM introduced that this design is to integrate the city landscape and the best view into the project; the building is thus known as the "Pazhou Eye". The twin towers feature special geometric shape and the hollow design of the top. The skyline therefore varies according to the heights of and the spaces between the buildings, and the whole view is a perfect blending of history and modernism.

会通
Conn-ectiv-ity

2001年《广州城市建设总体战略概念规划纲要》中提出"南拓、北优、东进、西联"的空间发展战略,2007年又通过《广州2020:城市总体发展战略规划》,广州逐渐向多中心组团式网络型城市格局转变,逐步构建起以空港、海港、铁路枢纽为龙头,以高快速道路和快速轨道交通"双快"交通体系为骨干的交通运输网络,引导未来城市的空间拓展。

在这样的背景下,广州新白云国际机场、广州南站、广州东站改造等项目相继完成,陆续建设起如广州科学城、广州大学城等教育科研办公建筑,不断提升城市的品质和竞争力,使空间格局和基础设施配置与城市的经济社会飞速发展相匹配。

在对外交流上,除了有完善的交通体系,会展建筑也起了很重要的作用。早在1957年起中国进出口商品交易会就在广州举办,其先后使用的各个展馆一路以来见证了广交会的变迁。

交通枢纽、会展中心,这些代表广州门户的建筑,承载着广州人的梦想,向世界展示着广州快速发展的步伐和永不停歇的脚步,宣扬着勇于开拓、开放包容的广州精神。

In 2001, the spatial development strategy of "southward expansion, northward optimization, eastward extension and westward combination" was proposed in *the Outline of General Strategic Conceptual Plan of the Guangzhou City Construction. In 2007, another report, namely Guangzhou 2020: General Urban Strategic Planning* was approved. Guangzhou has gradually been transformed to a city with a multi-center network and has built up a transportation network led by the airport, harbor and railway hub and then further enhanced by highway, expressway and rail rapid transit. All of them are aimed to lead the spatial expansion of the city in the future.

Under such a context, projects including New Guangzhou Baiyun International Airport, Guangzhou South Railway Station and Guangzhou East Railway Station have been completed one after another. All of them are designed to improve the quality of life and the competitiveness of the city and to make the spatial pattern and the infrastructure allocation be consistent with the city's rapid social and economic development.

In addition to a well-developed transportation system, buildings for convention and exhibition also play an important role in promoting international communication. As early as in 1957, the China Import and Export Fair was held in Guangzhou. The various pavilions used have witnessed the development of this Fair. Guangzhou Baiyun International Convention Center and Guangzhou Urban Planning Exhibition Center are functioning as the living room of Guangzhou for receiving guests and friends from all over the world.

Both the transportation hub and the convention and exhibition center represent the landmarks in Guangzhou, carry the dreams of its people and display Guangzhou's fast-growing pace and the spirit of exploration and inclusiveness.

承载历史辉煌，怀揣时代梦想。广州坚持以人为核心的城市发展理念，提升城市品质和国际竞争力，为实现中华民族伟大复兴的中国梦增添广州魅力。

Bearing the rich history and the dream of the new era, Guangzhou adheres to the human-centered urban development concept. It is dedicated to improving its urban quality and international competitiveness and to contributing to the realization of the great rejuvenation of the Chinese nation.

1978

- 1978 国务院利用外资建设旅游饭店领导小组决定在广州建设3座五星级宾馆
- 1978 改革开放城市重心逐渐东移
- 1981 全国第一个超级商场——广州友谊公司超级自选商场开业
- 1982 广州入选第一批国家历史文化名城
- 1982 广州图书馆开馆
- 1983–1985 三大中外合资宾馆先后建成开业
- 1984 广州经济技术开发区成立
- 1985 天河体育中心兴建
- 1987 第六届全国运动会在广州召开
- 1987 广州奥林匹克体育馆动工兴建
- 1983 白天鹅宾馆 White Swan Hotel
- 1984 中国大酒店 China Hotel
- 1984 花园酒店 The Garden Hotel
- 1987 天河体育中心 Tianhe Sports Center
- 1988 西汉南越王墓博物馆 Museum of the Mausoleum of the Nanyue King

1988

- 1990 地铁首期工程获批，全长12.7公里
- 1993 开建建设珠江新城
- 1994 正佳L场奠基
- 1996 作为内地最早一批商城的天河城开业
- 1997 广州地铁开通第一条线路
- 1997 广东公路主枢纽首个大型公用客运站——天河客运站建成
- 1998 广东奥林匹克体育馆动工兴建
- 1999 天河城广场全面开业
- 1999 提出珠江新城轴线规划"北优、东进"
- 2000 启动《广州城市建设总体战略概念规划纲要》，提出"南拓、北优、东进、西联"八字方针
- 2001 第九届全国运动会在广州举办
- 2001 《广州城市建设总体战略概念规划纲要》通过
- 2002 通过并正式实施南沙开发区规划
- 2004 广州成功获得2010年亚运会的举办权
- 2004 珠江新城轴线四大公建计划启动
- 2004 广交会正式移师琶洲会展中心
- 2005 国务院批复《广州市城市总体规划(2001—2010年)》
- 2005 广州地铁3号线开通
- 2005 正佳L场、太白汇、万菱汇等一系列大型商城相继建成开业
- 2006 《总规》修编，在八字方针基础上增加"中调"战略
- 2007 《L至2020》城市总体发展战略规划开展
- 1996 天河城 Teemall
- 1997 中信L场 CITIC Plaza
- 1997 广东美术馆 Guangdong Art Gallery
- 2001 广东奥体中心 Guangdong Olympic Sports Center
- 2001 广州体育馆 Guangzhou Gymnasium
- 2001 广州新体育馆 Guangzhou International Convention and Exhibition Center
- 2002 琶洲国际会展中心 Guangzhou International Convention and Exhibition Center
- 2003 逸夫人文馆 Shaw Building of Humanities
- 2004 广州白云国际机场 Guangzhou Baiyun International Airport
- 2005 正佳L场 Grandview Mall
- 2006 保利国际广场 Poly International Plaza

2008

- 2007 白云国际会议中心 Guangzhou Baiyun International Convention Center
- 2008 广东科学中心 Guangdong Science Center
- 2010 广州国际金融中心(西塔) Guangzhou International Finance Center (West Tower)
- 2010 广州塔 Canton Tower

中国出口商品交易会旧址（海珠广场）
The former site of China Export Fair (Haizhu Square)

中国出口商品交易会会址（流花路）
The site of China Export Fair (Liuhua Road)

1957年首届广交会现场
The 1st Canton Fair in 1957

1997年4月德国列车进入黄埔港
The train traveling from Germany to the Huangpu Port in April 1997

琶洲国际会展中心
The Guangzhou International Convention and Exhibition Center

白云机场二期改造工程
The Phase II reconstruction project of the Baiyun Airport

广州南站和谐号试运行
The Guangzhou South Railway Station and the trial run of China Railway High-speed (CRH) rail

2018

2017 《深化粤港澳合作 推进大湾区建设框架协议》签署

2016 作为天河路商圈收官之作的天环广场正式开业

2015 提出打造三大枢纽、完成一江两岸三带的目标

2014 《广州市历史文化名城保护规划》获批复

2011 国务院正式批复《广州市土地利用总体规划（2006—2020年）》

2010 首届亚洲残疾人运动会在广州开幕
2010 第16届亚洲运动会在广州开幕
2010 广州塔亮灯

2009 粤剧列入世界非物质文化遗产名录

2018 广州城市规划展览馆 Guangzhou Urban Planning Exhibition Center
2017 保利天幕广场（琶洲眼）Poly Skyline Square (Pazhou Eye)
2016 周大福金融中心（东塔）Chow Tai Fook Financial Center (East Tower)
2015 天环广场 Parc Central
2015 越秀金融大厦 Yuexiu Financial Tower
2013 珠江城（烟草大厦）Pearl River Tower (Tobacco Building)
2012 W酒店 W Hotel
2012 广州新图书馆 New Guangzhou Library
2012 富力盈凯广场 R&F Yingkai Square
2012 利通大厦 Leatop Plaza
2011 广州太古汇 Taikoo Hui
2011 广晟国际大厦 The Pinnacle (Guangzhou)
2010 广州南站 Guangzhou South Railway Station
2010 南越王宫博物馆 Archaeological Site Museum of Nanyue Palace
2010 广东省博物馆 Guangdong Museum
2010 广州大剧院 Guangzhou Opera House
2010 南沙体育馆 Nansha Gymnasium
2010 广州国际体育演艺中心 Guangzhou International Sports Arena
2010 广州亚运城综合体育馆 Asian Games Town Gymnasium
2010 广东奥体中心游泳跳水馆 Guangdong Olympic Sports Center Swimming and Diving Hall
2010 海心沙亚运公园 Haixinsha Asian Games Theme Park
2010 广州圣丰广场 Guangzhou Shengfeng Plaza

会通
Connectivity

琶洲国际会展中心
Guangzhou International Convention and Exhibition Center

地址：广州市阅江中路380号	Address: No.380 Yuejiang Middle Road, Guangzhou
建成时间：2002年	Completed in: 2002
占地面积：41.4万平方米（首期）	Floor area: 414,000m² (first phase)
总建筑面积：110万平方米	Total building area: 1,100,000 m²

2008年，广交会举办地整体转移到琶洲会展中心，确立了琶洲地区的发展目标为以会展为核心，以国际商务、信息交流、高新技术研发、旅游服务为主导，兼具高品质居住生活功能的RBD（休闲商务区）生态型城市副中心。琶洲国际会展中心位于琶洲岛上最核心位置，现已发展成为亚洲规模最大、世界规模第二的展贸中心，并随着历届堪称"世界第一展"的广交会的召开而蜚声于世、声名远扬。

会展中心采取"北低逐渐南高"的流线型设计，跌宕起伏、回转灵动的外观将一座巨大体量的建筑处理得轻盈、飘逸，极具音乐美感。建筑始于自然、融于自然、以人为本，集建筑艺术与现代科技于一身，是高科技、智能化、生态化完美结合的现代建筑。

2014年展馆四期建设完成后，展览面积逾50万平方米，成为世界上最大的展馆。未来，琶洲国际会展中心将继续代表着广州最先进的生产贸易水准向全世界展示广州的全新风貌。

Situated in the core area of the Pazhou Island, the Guangzhou International Convention and Exhibition Center has become Asian's largest and the world's second largest trade and exhibition center. The Center is gaining its reputation for the regular convening of the Canton Fair which is known as the world's largest exhibition. The height of the building gradually rises from the north to the south. This streamline shape has made the giant building light in the style and fluid in the form. Covering about 500,000 square meters after the completion of its fourth phase in 2014, it is the largest exhibition hall in the world. In the future, this Center will continue to represent Guangzhou's most advanced manufacturing and trade level and show the new look of Guangzhou to the world.

会通
Connectivity

广州白云国际机场
Guangzhou Baiyun International Airport

地址：广州市花都区
建成时间：2004年
占地面积：18平方公里
总建筑面积：1号航站楼52.3万平方米，
2号航站楼88.07万平方米

Address: Huadu District, Guangzhou
Completed in: 2004
Floor area: 18 km^2
Total building area: T1 is 523,000 m^2;
T2 is 880,700 m^2.

广州白云国际机场的前身是白云机场，位于白云山西边，1932年建成，用作军事用途。1963年改成民用机场后，经过多次扩建，仍不能满足日益增长的旅客吞吐量，于是1992年开始进行新机场的选址工作。经过多年准备，2004年旧机场停用，位于花都区的新白云国际机场T1航站楼正式启用。2012年机场扩建，2018年4月T2航站楼正式投入使用。

新白云国际机场的设计风格现代、简洁、流畅、精致，设计师以"白云"为设计主题，建筑的造型和空间设计都表达出"云端漫步、行云流水"的动感，表达出"轻盈、飘浮、流动"的感觉，与机场航站楼的功能特征非常吻合。T1与T2航站楼和谐一致，共同构建了"双子星"航站楼的完整形象，弧形的主楼、人字形柱、张拉膜雨棚及屋面花园等设计，都体现了岭南的地域特色。

新白云国际机场覆盖全球210多个通航点，其中国际及地区航点超过90个，通达全球40多个国家和地区。T2航站楼与国内、东南亚主要城市形成"4小时航空交通圈"，与全球主要城市形成"12小时航空交通圈"，被誉为"世界级巨无霸枢纽"。

Baiyun Airport, the predecessor of Guangzhou Baiyun International Airport, was built in 1932 and officially shut down in 2004 after over seventy years of operation. The New Baiyun International Airport is located in Huadu District. The T1 was officially opened on 5 August 2004, and the T2 came into service in April 2018. It is one of the largest and fully functional civil aviation hub airports in China. Its establishment has promoted the all-round development of Guangzhou's international airline market and formed a characteristic urban area with centralized industry and optimized resources, which has helped lay a foundation for Guangzhou to develop as an attractive modern city with strong influence on neighboring area.

一域商都新韵
New Architecture

白云国际会议中心
Guangzhou Baiyun International Convention Center

地址：广州市白云大道南1039-1045号
建成时间：2007年
占地面积：25.04万平方米
总建筑面积：31.6万平方米

Address: No.1039-1045 Baiyun Avenue South, Guangzhou
Completed in: 2007
Floor area: 250,400m^2
Total building area: 316,000 m^2

　　白云国际会议中心位于广州白云新城核心地段，占地面积27万平方米，是集会议、展览、酒店、宴会、演出、物业于一身的大型综合性会议中心。它拥有66间大小各异、灵活多变的专业标准会议厅，总面积超4万平方米，可满足大、中、小型各类会议使用需求，还拥有1079间客房。

　　白云国际会议中心摆脱了我们常见的孤立建筑的模式，用全新的设计理念描绘了一幅"白云织锦"。建筑师为减少对美好景观的破坏，并创造良好的建筑形式与其相适应，选择了极简的形式、抽象的特征、极端的直线、平坦的表面、常用的几何形体来设计建筑物。白云国际会议中心建筑外墙运用地方特色材料红砂岩，沉稳厚实，与白云山相映成趣。

　　在2008年巴塞罗那世界建筑节上，广州白云国际会议中心以其打破该类建筑固有的超大尺度，并把自然景观彻底引入的大胆构想在激烈竞争中脱颖而出，摘取公共类建筑的最高荣誉。

The Guangzhou Baiyun International Convention Center is located in the center of Baiyun New Town. In order to harmonize with the beautiful landscape and to create a matching architectural form, the architects adopted the simplest form, abstract elements, straight lines, flat surfaces and commonly used geometric shapes in the design. At the 2008 World Architecture Festival in Barcelona, the Guangzhou Baiyun International Convention Center, for its bold idea of breaking the conventions and incorporating the natural landscape, stood out from the fierce competition and won the highest award for public buildings.

会通
Connectivity

广州南站
Guangzhou South Railway Station

地址：广州市番禺区钟村镇石壁街道南站北路
建成时间：2010年
占地面积：61.5万平方米
建筑面积：48.6万平方米

Address: Nanzhan North Road, Shibi Street, Zhongcun Town, Panyu District, Guangzhou
Completed in: 2010
Floor area: 615,000m^2
Building area: 486,000m^2

广州南站是一个大型现代化铁路客运站，是华南地区最大、最繁忙的高铁站，是粤港澳大湾区、泛珠江三角洲地区的铁路核心车站，是连接京广高速铁路、广深港高速铁路、贵广高速铁路、南广铁路和广珠城轨及北京、上海、武汉、深圳和香港的世界级综合交通枢纽，与广州站、广州东站和广州北站组成全国四大铁路客运枢纽（京沪穗汉）之一——广州铁路客运枢纽。

车站主要是钢结构，总用钢量7.9万吨，外观设计独具岭南特色，巨大的玻璃穹顶形似6片飘浮于空中的芭蕉叶，其中部为64米大跨拱形结构。站房屋顶以一片片的"芭蕉叶"为基本单元，通过中央采光带的串联，形成极具特色的建筑形态。

广州南站经停车次数量位列全国第一。截至2010年1月，广州南站共设15站台、28站台面、28股道。广州南站以行车时速可达300公里以上的高铁，实现了珠三角主要城市之间1小时内互通往来。目前，其始发高铁（动车）可直达国内18个省、自治区、直辖市，4小时内覆盖粤、桂、滇、黔、湘、鄂6省主要城市，8小时内直达郑州、北京、西安等城市。

The Guangzhou South Railway Station is a large modern railway station located in Panyu District in Guangzhou. Serving as the South Gate of Guangzhou, it stands as the largest and busiest high-speed railway station in South China, functioning as a world-class comprehensive transportation hub that connects Beijing, Shanghai, Wuhan, Shenzhen, and Hong Kong. Guangzhou South Railway Station, along with Guangzhou Railway Station, Guangzhou East Railway Station and Guangzhou North Railway Station, form four major railway passenger transportation hubs in China. The station adopts a steel structure, consuming 79,000 tons of steel in total. The exterior of the station features unique Lingnan style. The roof is in the form of "banana leaves". Through the connection of the central lighting belt, it forms a distinctive architectural appearance.

会通
Connectivity

广州城市规划展览馆
Guangzhou Urban Planning Exhibition Center

地址：广州市白云区展览路1号
建成时间：2018 年
占地面积：35333.8平方米
建筑面积：8.4万平方米

Address: No.1 Zhanlan Road, Baiyun District, Guangzhou
Completed in: 2018
Floor area: 35333.8 m^2
Building area: 84,000 m^2

广州市城市规划展览馆位于白云山旁，白云新城核心区域，是一座承载广州时代精神、城市品质的新地标。

展览馆由中国工程院院士何镜堂亲自领衔设计。建筑设计以地域建筑特色为依托，采用模仿岭南民居蚝壳墙质感的深灰色为主调，通过下沉庭院打造岭南水乡风情，通过骑楼元素展现传统韵味。馆内展陈融合了传统与现代、文化与科技，富有艺术性地展示了广州传承千年的文化底蕴和城市发展的前世今生，成为生动展现广州城市规划建设历史、现在、未来的"城市文化客厅"和展示广州城市之光的重要场所。

展览馆凭借功能复合、空间渗透等先进的设计理念与创新技术，将文化、建筑艺术和功能完美结合，打造城市规划展览区、综合会展区和流动空间等多维空间。馆内拥有展示广州7434平方公里的全域物理沙盘，4D、5D影片，大型航拍图等，是目前国内一流的复合型规划展览馆。

会通
Connectivity

The Guangzhou Urban Planning Exhibition Center is situated at the heart of Baiyun New Town by the Baiyun Mountain. It is a new landmark exhibiting the spirit of the new era and the essence of the city. This exhibition center is designed by HE Jingtang, an academician of the Chinese Academy of Engineering. The interior design incorporates both traditional and modern elements, and integrates classical culture with technologies. It skillfully displays Guangzhou's thousands of years of cultural history and tells the story of the city's development. It is like a living room showing how Guangzhou has been developed from the past and how it will be planned in the future, and is also an important venue presenting the beauty of this city.

项目导览
Project navigation

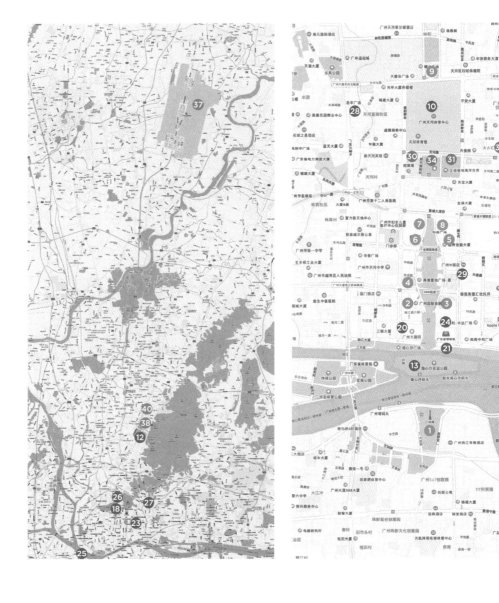

项目导览
Project Navigation

1. 广州塔
   Canton Tower
2. 广州国际金融中心(西塔)
   Guangzhou International Finance Center (West Tower)
3. 周大福金融中心(东塔)
   Chow Tai Fook Financial Center (East Tower)
4. 富力盈凯广场
   R&F Yingkai Square
5. 越秀金融大厦
   Yuexiu Financial Tower
6. 珠江城大厦(烟草大厦)
   Pearl River Tower (Tobacco Building)
7. 广晟国际大厦
   The Pinnacle (Guangzhou)
8. 利通大厦
   Leatop Plaza
9. 中信广场
   CITIC Plaza
10. 天河体育中心
    Tianhe Sports Center
11. 广东奥体中心
    Guangdong Olympic Sports Center
12. 广州新体育馆
    Guangzhou Gymnasium
13. 海心沙亚运公园
    Haixinsha Asian Games Theme Park
14. 广东奥林匹克体育中心游泳跳水馆
    Guangdong Olympic Sports Center Swimming and Diving Hall
15. 广州亚运城综合体育馆
    Asian Games Town Gymnasium
16. 广州国际体育演艺中心
    Guangzhou International Sports Arena
17. 南沙体育馆
    Nansha Gymnasium
18. 西汉南越王墓博物馆
    Museum of the Mausoleum of the Nanyue King
19. 逸夫人文馆
    Shaw Building of Humanities
20. 广东科学中心
    Guangdong Science Center
21. 广州大剧院
    Guangzhou Opera House
22. 广东省博物馆
    Guangdong Museum
23. 南越王宫博物馆
    Archaeological Site Museum of Nanyue Palace
24. 广州新图书馆
    New Guangzhou Library
25. 白天鹅宾馆
    White Swan Hotel
26. 中国大酒店
    China Hotel
27. 花园酒店
    The Garden Hotel
28. 广州圣丰广场
    Guangzhou Shengfeng Plaza
29. W酒店
    W Hotel

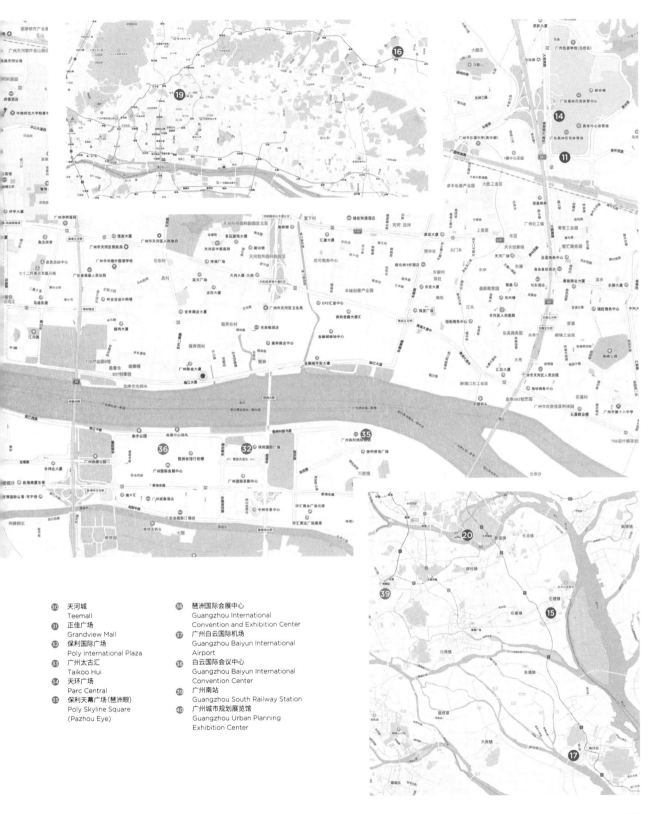

- ㉚ 天河城
  Teemall
- ㉛ 正佳广场
  Grandview Mall
- ㉜ 保利国际广场
  Poly International Plaza
- ㉝ 广州太古汇
  Taikoo Hui
- ㉞ 天环广场
  Parc Central
- ㉟ 保利天幕广场(琶洲眼)
  Poly Skyline Square (Pazhou Eye)
- ㊱ 琶洲国际会展中心
  Guangzhou International Convention and Exhibition Center
- ㊲ 广州白云国际机场
  Guangzhou Baiyun International Airport
- ㊳ 白云国际会议中心
  Guangzhou Baiyun International Convention Center
- ㊴ 广州南站
  Guangzhou South Railway Station
- ㊵ 广州城市规划展览馆
  Guangzhou Urban Planning Exhibition Center

作者名录
Author List

本书收录了众多机构和摄影师提供的精彩的图片，在此表示感谢。图片的版权归属为图片的提供者。

This book contains wonderful pictures provided by many institutions and photographers, to whom we express our thanks. The copyright of the images belongs to the providers of the images.

广州市设计院：
P4-2 P15 P20 P22 P30 P31-1/2 P48 P49 P51-1/2/3 P58-1/2 P57 P59
P68-2/3 P69 P92 P93-1/2/3 P94 P95 P96 P97-1 P102 P103 P106
P120 P126 P127-1 P129 P130 P131-1/2/3 P138

DUO建筑影像：
P9 P4-3 P5-1/2 P24 P25 P32 P33 P40 P41 P42 P63-1/2 P76 P77-1/2/3
P87-1 P88 P89-2 P97-2 P108 P109 P113-2 P121 P115-1/2 p121 P124
P125 P127-2 P136-2/3 P156 P157-1/2/3

华南理工大学建筑设计研究院：
P1 P5-3 P17-1/2 P18-1/2/3 P19 P26 P27 P28 P29-1/2
P37 P38 P39-1/2 P53-1/2 P54 P112 P113-1
P78-1/2/3 P79 P80 P144-1/2 P145 P147 P165

广东省建筑设计研究院：
P66 P67-1/2 P89-1 P143-6 P148-1/2 P149 P150

广州市规划院：
P60 P61-1/2 P62

广州市档案馆：
P11-3 P47-4/5/8 P52 P101-1 P143-1/5 P153 P154 P155

孙一民工作室：
P64 P65-1/2/3 P70-1/2 P71

冼剑雄事务所：
P34 P35-1/2/3

许李严建筑设计有限公司：
P114 P115-1/3

太古汇（广州）发展有限公司：
P132 P133-1/2 P134 P135-1/2

覃光辉 P13-1 P68-1
李波 P2 P6 P56
苏俊杰 P4
董才章 P12
陈靖文 P13-1/2
李树竞 P14
杨艺 P16
曾汉鹏 P1 P21
陈忧子 P36
陈冲 P50
黄爱萍 P75-3
郑裕彤 P82
陈兆武 P83
李志强 P84-1
张广源（中国建筑设计研究院）P84-2 P85 P86 P87-2
陈展荣 P90
杨艺 P104
陈汉添 P107
刘达明 P110
李志颖 P119-6
尹金山 P122
孟俊锋 P128
张永林 P136 P152
冯健文 P139
何进 P146
战长恒 P158-1/2/3/4 P159 P161 P163

图片来自网络：
P11-1/2/4/6
P47-1/2/3/6/7
P75-1/4/6/7/8
P101-2/3/4
P119-1/2/3/4/5
P143-2/3/4/7

广州建筑图册
广州滨水建筑图册

WATERSIDE LANDSCAPE

一江风景图画

广州市规划和自然资源局 编

花城出版社
中国·广州

图书在版编目（CIP）数据

广州建筑图册. 广州滨水建筑图册 / 广州市规划和自然资源局编. -- 广州：花城出版社，2023.12
ISBN 978-7-5749-0098-1

Ⅰ．①广… Ⅱ．①广… Ⅲ．①建筑文化－广州－图集 Ⅳ．①TU-092

中国国家版本馆CIP数据核字(2023)第254876号

主编单位：广州市规划和自然资源局
承编单位：广州市城市规划设计有限公司
　　　　　广州市设计院集团有限公司

出 版 人：张　懿
责任编辑：陈诗泳
责任校对：衣　然
技术编辑：林佳莹
装帧设计：广州市耳文广告有限责任公司

书　　名	广州建筑图册·广州滨水建筑图册 GUANGZHOU JIANZHU TUCE·GUANGZHOU BINSHUI JIANZHU TUCE
出版发行	花城出版社 （广州市环市东路水荫路11号）
经　　销	全国新华书店
印　　刷	佛山市迎高彩印有限公司 （佛山市顺德区陈村镇广隆工业区兴业七路9号）
开　　本	787毫米×1092毫米　16开
印　　张	10　1插页
字　　数	96,000字
版　　次	2023年12月第1版　2023年12月第1次印刷
定　　价	288.00元（全三册）

如发现印装质量问题，请直接与印刷厂联系调换。
购书热线：020－37604658　37602954
花城出版社网站：http://www.fcph.com.cn

广州的建筑，是东西交流、南北对话、新旧共融的集大成者

南沙天后宫	琶洲塔	赤岗塔	莲花塔
Nansha Tianhou Temple	Pazhou Pagodas	Chigang Pagodas	Lotus Pagoda
明代	明—清	明—清	明—清

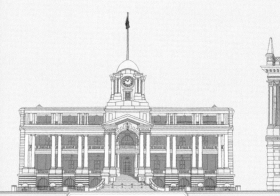 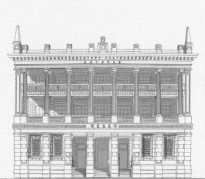  

粤海关旧址	广东邮务管理局旧址	南方大厦	塔影楼
Site of Former Guangdong Customs	Site of Former Guangdong Post Administration	Nanfang Building	Taying Building
1916年	1916年	1916年	1916年

广州白天鹅宾馆	星海音乐厅	广东美术馆
White Swan Hotel	Xinghai Concert Hall	Guangdong Museum of Art
1983年	1998年	1997年

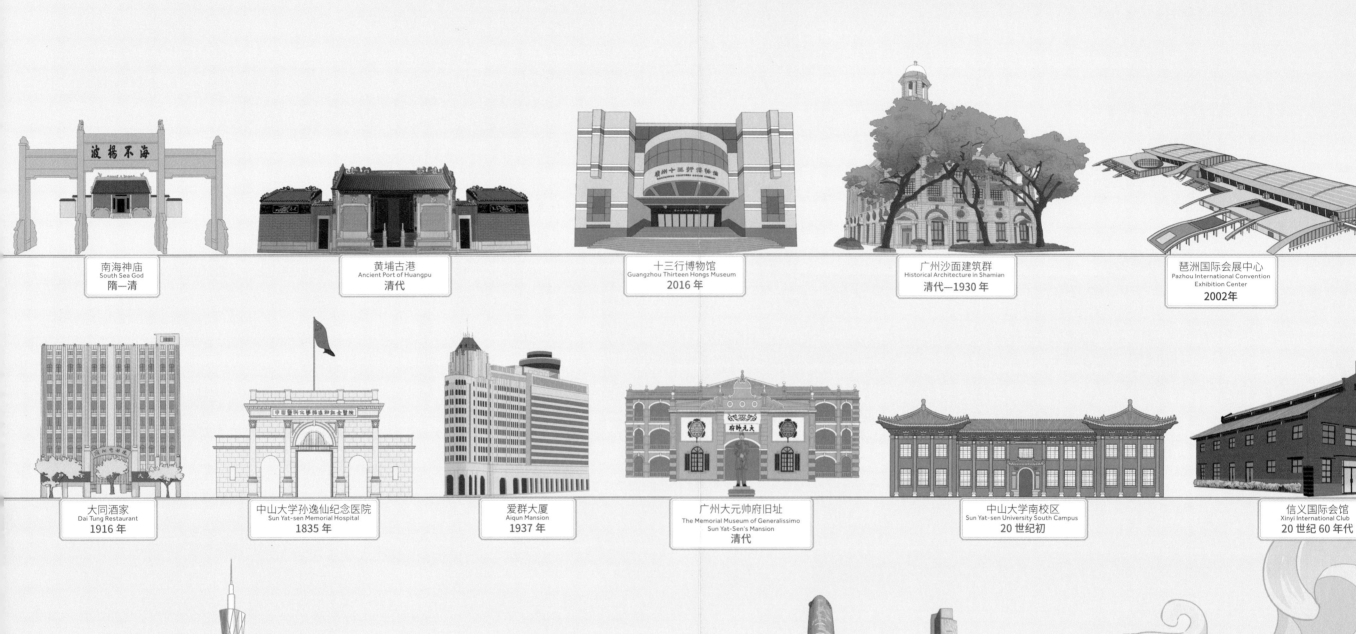

写在前面
The Editor's Notes

序章 | Prologue

广州兴城，得益于水。这座比威尼斯大17倍的特大城市，江河湖海俱备，水滋养着这方水土，水哺育着此处黎民，因水兴航运，借水溉良田，依水筑村落，凭水生凉风。水，是广州的魂与骨。

今天，广州城的滨水兴城计划，以更磅礴壮丽的格局展开，一个世界级的滨水城市，正勃然崛起，本书用四个章节徐徐展开一幅广州的滨水生活图卷：

《江岸》沿母亲河珠江，有看不厌的风景——3段各10公里珠江江岸线，西边是19世纪初的长堤商业圈，东边是鳞次栉比的珠江新城现代化楼宇，从西向东，按广州发展的时间轴向前铺开。

《海丝》广州是由江和海合力催生而成的一片传奇土地，早在汉唐之前，广州便借临海地利，与各国的商贸往来不断；本章从海上丝绸之路的新旧痕迹出发，概述航运的便利给广州带来的机遇。

《湖光》新中国成立后，为了蓄洪排涝、营造城市景观，兴建了四大人工湖。今天，更多翡翠般的湖泊布列于城市各处，和珠江各支流联结成一体，形成了丰富多元的水系，为城市增添了美丽的肌理。

《栖水》广州境内河网密布，人们因地制宜，驭水为通途和良居，依水诗意而居。旧时的塱头村改水泽为良居，大岭村引水绕村、围海造田；现代广州人疏浚涌水，将污水化为清流，享现代化便利，又兼保文脉绵延，处处都是广州人驭水的典范。

水利百物，水养育的广州城将以水之包容与博大精神，迎接新的机遇与挑战。

The *Riverfront Landscape and Attractions* presents the three 10-km riverfront landscape belts along the Pearl River from the west to the east, which is also a time line that epitomizes the evolution of the city.

The *Maritime Silk Road* briefs the opportunities the convenient shipping network brought to the city as evidenced by the heritage of the Maritime Silk Road in the city.

The *Lake Parks* introduces the lake parks in the city. Together with the Pearl River tributaries, they create a diverse water system and attractive ecological environment for the city.

Waterfront Living portrays the poetic waterfront life in Guangzhou, where people harnessed the abundant water resources for convenient transportation and poetic life.

目录
Contents

序章　Prologue

水绕城成画　A Picturesque Waterfront City	6
水连城，桥连家　Connection with Water forms Basis of The Urban Life	13
牧歌与礼赞，乡愁与展望　Heritage preservation vs Urban development	20
这座以水为骨的城，比威尼斯大17倍　This City Born Out of Water Is 17 Times Larger Than Venice	22
海孕育的大地，江灌溉的沃土　A Land Reclaimed from the Sea and Flourishing Over the River	25
城的年华，水的模样　Water in Local History and Tradition	27
驭水为良田，依水为良居　Harness Water to Develop Fertile Farmlands and Poetic Habitats	28
当时间沉淀成沙洲　Blessed Land vs Artificial Island	31

第一章：江岸　Chapter 1: Riverfront Landscape and Attractions

江岸如画　Pastoral Songs and Praise	34
粤海关旧址　Site of Former Guangdong Customs	39
广东邮务管理局旧址　Site of Former Guangdong Post Administration	41
南方大厦｜大同酒家｜塔影楼　Nanfang Building / Dai Tung Restaurant / Taying Building	43
爱群大厦｜中山大学孙逸仙纪念医院　Aiqun Mansion / Sun Yat-sen Memorial Hospital	45
大元帅府旧址　Memorial Museum of Generalissimo Sun Yat-sen's Mansion	47
中山大学南校区　Sun Yat-sen University South Campus	49
信义国际会馆　Xinyi International Club	50
太古仓　Pacific Warehouse	51
琶醍　Party Pier	53
红专厂　Redtory	53
广州白天鹅宾馆　White Swan Hotel	55
星海音乐厅　Xinghai Concert Hall	57
广东美术馆　Guangdong Museum of Art	58
广东华侨博物馆　Guangdong Museum of Chinese Nationals Residing Abroad	59
广州塔｜海心沙　Canton Tower / Haixinsha Island	63
广州大剧院　Guangzhou Opera House	65
广州国际金融中心｜广州周大福金融中心　Guangzhou International Finance Center / CTF Finance Center	67
广东省博物馆　Guangdong Museum	69
广州图书馆　Guangzhou Library	73

目录
Directory

第二章：海丝
Chapter 2: Maritime Silk Road

南沙天后宫 Nansha Tianhou Temple	77
琶洲塔ǀ赤岗塔ǀ莲花塔 Pazhou Pagoda / Chigang Pagoda / Lotus Pagoda	79
南海神庙 South Sea God Temple	83
黄埔古港 Ancient Port of Huangpu	85
十三行博物馆 Guangzhou Thirteen Hongs Museum	87
广州沙面建筑群 Historical Architecture in Shamian	88
琶洲国际会展中心 Pazhou International Convention Exhibition Center	93

第三章：湖光
Chapter 3: Lake Parks

麓湖公园 Luhu Park	101
流花湖公园 Liuhua Lake Park	103
东山湖公园 Dongshan Lake Park	104
荔湾湖公园 Lychee Lake Park	105
海珠国家湿地公园 Haizhu National Wetland Park	107
白云湖公园ǀ南湖游乐园 Baiyun Lake Park / Nanhu Lake Amusement Park	108

第四章：栖水
Chapter 4: Waterfront Living

沙湾古镇 Shawan Ancient Town	113
大岭村 Daling Village	117
塱头古村 Langtou Village	121
聚龙村 Julong Village	125
余荫山房 Yuyin Mountain House	127
宝墨园 Baomo Garden	129
小洲村 Xiaozhou Village	131
云桂桥ǀ利济桥ǀ汇津桥 Yungui Bridge / Liji Bridge / Huijin Bridge	133
龙津桥ǀ石井桥ǀ五眼桥 Longjin Bridge / Shijing Bridge / Wuyan Bridge	135
东濠涌 Donghao Canal	139
荔枝湾涌 Lychee Bay	141
粤剧艺术博物馆 Cantonese Opera Art Museum	143
猎德涌 Liede Canal	145

水绕城成画
A Picturesque Waterfront City

珠江水滋养的广州城,四时草木葱茏、花朵竞芳,四方江水丰沛、波光云影……因为水,宜居的城,成一卷卷悦目的画。

水绕城成画
A Picturesque Waterfront City

珠水环抱的广州城，四方皆美，江河奔涌壮丽，湖光映照云霞，因为水，钢筋铁骨的城，有了别样的柔波，有了动人的生机。

水绕城成画
A Picturesque Waterfront City

霞光笼罩城东的此刻，一城的人向家起程，跨过珠江，回到城的另一端，当夕阳跃入江中，两岸便亮起万家灯火，映在江波上，像星光。

东江水和增江水,自城东簇拥成洋洋江河奔向大海;
江之东,曾是田园牧歌里的乡野和村落,今天是高楼接天、灯火璀璨的新城;
从江之东走到江之西,
一路都是风景。

水连城，桥连家
Connection With water Forms
Basis of The Urban Life

北江水、西江水和流溪河，于城西汇集成绕城碧波；
珠江之西，曾是海丝路上声名显赫的码头，今天是新老交织、人情暖暖的老区；
从江之西走到江之东，
一桥桥都是时光。

东西绕城的珠江水，将一座广州城分成隔江而望的城南城北。跨过珠江，从前广州人叫"过海"。

海珠桥上推看自行车的人海、渡口排队坐渡轮"过海"的人龙，是每个老广脑海中的似水流年。

今天，在珠江上的桥，西到东、南到北，已多不胜数，城南城北、城东城西，早已处处通途。城的肌理，在水网中又织进路桥之网，繁华壮美、生机勃发。

1	2	4
	3	5

1. 华南大桥以东，是勃勃兴起的新区。
2. 游船通过海印桥和江湾大桥。
3. 解放桥和海珠桥的灯火，映在珠江的柔波里。
4. 海珠桥加木棉花，是老广们的乡愁标配。
5. 东沙桥从广州圆务飞越。

从城西的白鹅潭回望广州城,霞光笼罩这繁华之都,日间的奔忙与勤勉刚刚落画,老广们夜色中的人间烟火,又将开锣,一江两岸,满满都是热热闹闹的生活味道。

牧歌与礼赞，乡愁与展望
Heritage Preservation vs Urban Development

蓝与绿，海洋与绿洲，农耕文明与现代文明，
这水做的城，既古朴又摩登，处处是看不厌的肌理、拍不完的风光。

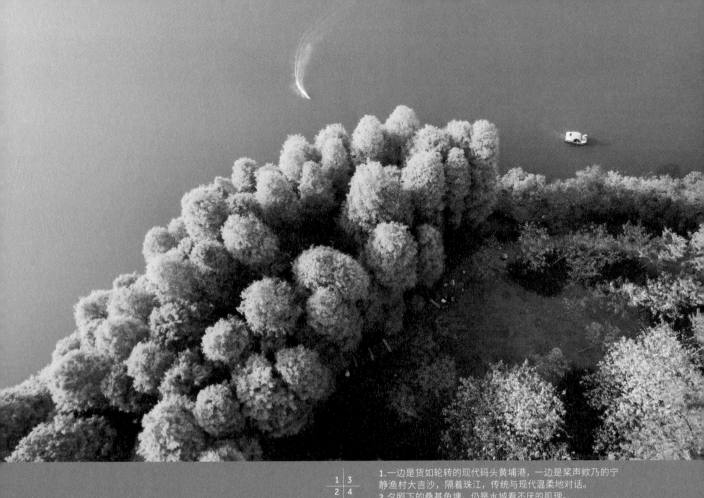

1. 一边是货如轮转的现代码头黄埔港，一边是桨声欸乃的宁静渔村大吉沙，隔着珠江，传统与现代温柔地对话。
2. 夕照下的桑基鱼塘，仍是水城看不厌的肌理。
3. 当年为了排内涝而开挖的麓湖，早已树木葱茏。
4. 珠江沿岸，暖暖的归家灯火。

这座以水为骨的城，比威尼斯大 17 倍

珠江——开放与融汇

　　弱水三千汇入广州，水利百物，广州便成四季如画的花城，成五谷丰登的穗城。

This City Born Out of Water Is 17 Times Larger Than Venice.

A city born out of water, Guangzhou is crisscrossed by a dense network of rivers and lakes. The Pearl River runs through the city in grandeur, while the Liuxi River, a tributary of the Pearl River, flows through the northern part of the city, forming the backbone of the city's water system.

1	2	3
4	5	6

1. 花都白坭水边的九曲画廊。
2. 车陂端午彩龙走亲。
3. 白鹅潭夜色。
4. 流溪河水库。
5. 黄埔港码头。
6. 珠江上的新航程。

"圳""陂""潭""埔""湾""涌""溪""滘""沙""洲",在广州,无数地名带着水的基因。

广州这座依水而生之城,江水环绕、河网密布、湖泊丰盈,东江、西江、北江,三江奔腾而来,洋洋壮阔,而清澈的流溪河、增江自北部的群山汇下,给珠江再注入柔波。水的馈赠无处不在,富含养分的流沙形成新的土地、水量丰沛的航道带来生生不竭的商机,广州人靠水吃水,依珠江围出桑基鱼塘,将泽地建设成富足安康的人居;又引珠水绕村,得灌溉、浣洗的便利,又使水通财通,保持与世界接连、与机遇接连。

历经多年积淀,广州,最终形成今天通山达海的壮阔底色。

七星岗 举世罕见的深入内陆的七星岗海蚀遗迹，印证着广州的海洋属性。

常春岩 南海神庙附近的南湾村，有一块20亩大的海蚀石，当年无数越洋商船便是从南海神庙前的狮子洋穿梭往来。

五仙观仙人拇迹 晋时的珠江岸边坡山古渡口，现在已离珠江岸1公里远，可以想象，曾经珠江江面之浩瀚宽阔。

千年古道遗址 广州自建城，2200多年来未移城址，而南汉以来，北京路便是从珠江登岸，进入广州城的主要官道。

黄埔古港 自宋开始，黄埔古港就是海上丝绸之路的重要一环，到了清代一口通商的80年间，逾5000艘外国商船在此停泊。

琶洲塔 海上丝绸之路的外国商船，看到珠江边的莲花塔、琶洲塔、赤岗塔，便知广州已在眼前。

琶洲国际会展中心 在当年海上丝绸之路的航标塔、风水塔琶洲塔附近，有目前亚洲最大的会展中心——琶洲国际会展中心。

南沙港 在广州与大海拥抱的最南端，是新兴的南沙自贸区，其港口商船如织，一派繁华之象。

海孕育的大地，
江灌溉的沃土

A Land Reclaimed from the Sea and Flourishing Over the River

80多年前中山大学教授吴尚时与学生曾昭璇在广州七星岗发现了古海岸遗址，举世罕见、深入内陆100公里的海蚀崖壁，揭开了6000多年前一卷海、江、大地相互角力、恢宏壮阔的画面。

然后，大海从白云山、西樵山下一路退去，留下绵延不断的沃土与良田。而数路江河，在海与江合力冲刷出的汊道里川流不息，所以此处的河流，千溪万涌、四通八达。

这片土地，也是世间罕有的2000年未易城址的古城。南湾村巨大的海蚀岩石常春岩，岩刻题字由明代大儒陈白沙所书；明代官制道场五仙观旁的坡山古渡遗迹——仙人拇迹，这四字则是白沙先生的学生、三部尚书湛若水所题——固然海上丝绸之路的贸易通道早早通航，文脉与商运在此地并举共荣，同样从未中断。

借大海接连七洲、凭河道通达四方，航运之便捷，为广州带来了长盛不衰的商机。全世界都知道这个曾经被叫Canton的地方，每一件运往世界的产品，都是精美的硬通货，当年一艘满载归航的远洋轮货值，抵得上一个国家的国内生产总值。

今天，海上丝绸之路的痕迹在广州无处不在，而格局更壮阔的新海上丝绸之路重新启程。广州这个海与江孕育出的宠儿，以更加通达与包容之心与世界连接，再次焕发绚丽的光芒。

The joint efforts of rivers and sea created the abundant canals and fertile land of Guangzhou. The convenient shipping network brought sustained business opportunities to the city and made Canton a reputed name worldwide. In old days, a fully-loaded ocean ship returning from Canton could be worth the GDP of a country.

Today, the heritage of the Maritime Silk Road can still be found at every corner of the city and we are to embark on a more ambitious journey on the Maritime Silk Road.

龙船脊 龙船脊这种广州常见的屋脊形制，映射出广州人对水的无比依恋。

龙船走亲 农村再怎么变，都会保留龙船走亲的仪式感。

河涌 基于交通、生活便利和排涝需求，河涌绕村是水乡的常态风貌。

桑基鱼塘 这种基于水的立体养殖模式绿色高效，至今仍意义非凡。

风水塘 祠堂前的风水塘，为族人聚财的同时也改善了小气候。

天后宫 近海人家信奉保佑远洋船只出入平安的天后娘娘。

南海神庙 受过皇家封礼的南海神，宋代就开始保佑商船来往。

北帝庙 司掌水的北帝，自然在水乡广州城很受推崇，所以水口位常见北帝庙。

城的年华，
水的模样

Water in Local History and Tradition

天时厚待加上人力勤勉，广州自成举足轻重的鱼米之乡、商贸重镇。人们靠水吃水，敬水畏水，倚水力贩运货物，借水流促生微风，水是广州人的工具，是倚仗，也是景观。

祠堂以龙船造型为脊，村落与村落之间以彩装龙船隆而重之互访，以水为纽带，各村互结亲谊；人们拜司水的龙母、北帝和南海神，以庇佑水波不兴、出入平安；又善用潮汐，引江河之水绕村，涤衣、灌溉，享水的种种便利；再依江、湖而筑围出连绵壮观的桑基鱼塘；还依江两岸开出商埠，面对海洋带来的机遇与挑战。

广州人，离不了水，广州城由水滋养，万物欣荣。

依水而造的建筑，宜商宜居，各有风姿，或恢宏如庙宇，或巍峨如宗祠，或富丽如华邸，每一处都有广州人的巧思与审美。

因水生、因水兴的广州，水是永恒不变的灵魂。

In old days, the ridges of the ancestral halls in Guangzhou were shaped to resemble dragon boats. The villages were interconnected with each other via canals, while villagers developed family ties and friendship based on such connections and visited each other by riding the elaborately dragon boats. They worshiped the Gods and Goddesses in charge of water, such as Dragon Mother, Pai Tei (The North Deity) and South Sea God to pray for safe voyage. They utilized the sea nater from the tides for various purposes in and around the village, built fascinating mulberry dykes and fish ponds beside the rivers and lakes, and developed trade ports along the river to benefit from the opportunities brought by the advancements of shipping.

Guangzhou is a city born from water and has flourished alongside it, while water remains the eternal soul of the city.

驭水为良田，依水为良居
依水而居的现实浪漫主义

广州境内河流密如蛛网，有西江、东江、北江流经，更有流溪河、增江等灌注，广州人善于驭水并化泽地为良居、造田、挖湖、兴鱼塘，造一方诗意家园。

Harness Water to Develop Fertile Farmlands and Poetic Habitats

Guangzhou has offered different versions of examples for rational water utilization. In Langtou village, a village built in the Yuan Dynasty, the swamp was transformed into attractive human habitat, where the houses were placed on high land in the center while low-lying ponds in the periphery were preventing waterlogging and theft. The Daling Village built in the Song Dynasty used the sea water from the tides to form meandering waterways around the village, creating a good Fengshui pattern.

In the Qing Dynasty, the wealthy families in the west of the city flocked at the New Litchi Bay to build luxurious mansions by the canal, while Ju brothers, the founders of the Lingnan Painting School, built the Ten Fragrance Garden at the waterfronts in the south of the city to lead the students to portraying the wonderful waterfront views.

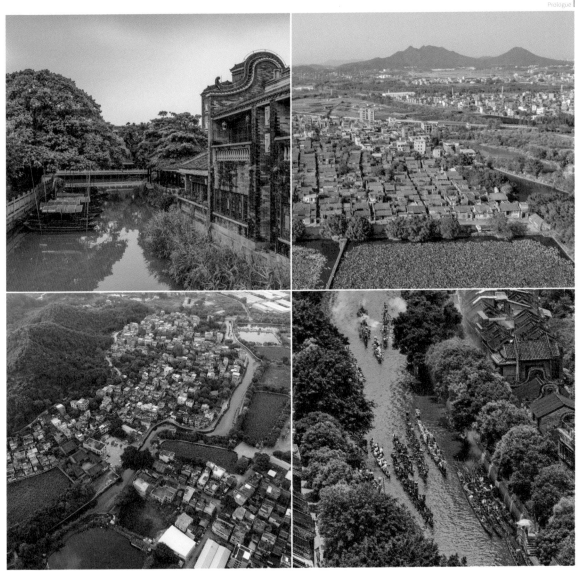

1. 升级改造后新荔枝湾涌。
2. 南沙东涌水乡。
3. 化泽地为良居的塱头村。
4. 大岭村水道与狮子洋相通。
5. 猎德涌十乡八里的老表走亲。

自宋代开始疏浚六脉渠，以水为纽带，打通全城命脉，一来可蓄水泄洪，二来又便于运输，三则水转运转，处处风门水口，皆成旺地。

这种对水合理化运用的思路，在广州的乡间也多有范本：元代建村的塱头村将湖泽化为良居——风水塘一字排开布列村前，人居较高，积水流向村四周水塘，防涝防盗，乡人在水中央安居乐业；宋代建村的大岭村，则引水绕村，靠山对水，祠堂前水道蜿蜒，风生水起。

到清代，六脉渠虽变化良多，但水仍是广州不变的肌理——城西，首富梯队云集新荔枝湾，依水建成华屋豪厦、美园仙馆；城南，岭南画派师祖居氏则依瑶溪开设十香园，带学生沿水画尽美景。

珠江水的柔波，每一朵都是生生不息的故事。

当时间沉淀成沙洲
/天赐的土壤，人造的绿岛/

珠江水携富含养分的沙土从各路汇入广州，历千万年沉积，成珠江里的一片片肥沃沙洲。

人们在这些沙洲上，垦荒造房，种花栽树，终将荒岛建成绿岛。

今天，这些沙洲的开发，实现了最大限度的公共性，优美的广场、设施先进的博物馆、图书馆、大剧院……全部敞开大门，是广州人献给这座城气势恢宏的大手笔。

而更多沙洲——大吉沙、海鸥岛、龙穴岛、观龙岛……完好地保存了以渔耕为底色的田园牧歌，并在蓝图中，将以更宜居更便利的规划，更好地成为广州人的心之憩所。

享现代之便利，拥传统之诗意，广州人，果然是有福气的。

Blessed Land vs Artificial Island

The fertile sand soil brought by the Pearl River water into the city formed a series of sandbanks in the Pearl River after thousands of years of sedimentation.

On these sandbanks, people built houses, grew crops and planted trees and flowers, eventually turning the deserted island into green oasis.

Today, the developments on these sandbars have helped the city achieve the greatest degree of public spaces. The beautiful square, state-of-the-art museum, library and the opera house are new landmarks that the Guangzhou people created for the city.

1	
2	
3	4

1. 珠江上的沙岛沙面。
2. 二沙岛和海心沙。
3. 珠江新东岸，未来可期。
4. 人工种植的红树林使海鸥岛形成丰富的小生态环境。

第一章：江岸
Chapter 1: Riverfront Landscape and Attractions

穿城而过的珠江水，也为广州带来了一江两岸百看不厌的美景。广州是中国最早拥有现代化江岸的城市，江岸白鹅潭、长堤、海珠广场、花城广场……这赏不完的30公里精品珠江景观带，也是一卷翻不完的广州城史。沿着滨水慢行路径漫溯：长堤是珠江景观带的核心，当年的长堤上，最现代化的、最奢华的摩天楼比肩而立，跨过一个世纪，今天仍是岁月的经典；今天的珠水东岸，新时代的传奇正在奋笔疾书；新广州老广州，沿着江岸交织，每一楼、每一宇，都是岁月砌成的歌。

Guangzhou is the first in China to develop the modern riverfront landscape. The 30km Pearl River landscape belt is a book of profound city history. Changdi used to be the core of the landscape belt in old days. The most modern top-grade tall buildings at that time were built one after another there and still remain classic masterpieces even today. Nowadays a legend of the new era is being unfolded on the east bank of the Pearl River.

江岸如画
Pastoral Songs and Praise

浩浩珠江水为2200多岁的广州城输送绵绵不绝的活力,仰仗着母亲河珠江,广州人沿江创造了读不完的传奇、看不够的风景——西起白鹅潭、东至南海神庙,西、中、东3段各10公里珠江沿岸,正代表着三个辉煌时期的最美时代风貌。

如织的商船已绕道珠江外航道通行，而广州最美的30公里景观带，和30公里的珠江柔波，是广州人献给这座两千年来活力无限的广州城最迷人的绿与蓝。

1	
2	3

1. 珠江新城景观带。
2. 白鹅潭。
3. 珠江后航道。

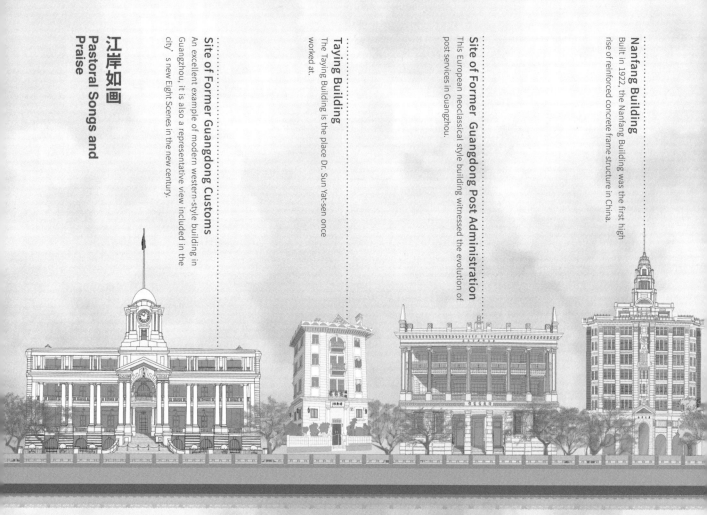

Nanfang Building
Built in 1922, the Nanfang Building was the first high rise of reinforced concrete frame structure in China.

Site of Former Guangdong Post Administration
This European neoclassical style building witnessed the evolution of post services in Guangzhou.

Taying Building
The Taying Building is the place Dr. Sun Yat-sen once worked at.

Site of Former Guangdong Customs
An excellent example of modern western-style building in Guangzhou, it is also a representative view included in the city's new Eight Scenes in the new century.

江岸如画
Pastoral Songs and Praise

粤海关旧址
该建筑被视为广州近代西洋建筑的典范，也是新世纪羊城新八景之一『珠水夜韵』的标志性景点。

塔影楼
塔影楼矗立江边，位于原联兴码头登船处，因倒影似塔，取名『塔影楼』。

广东邮务管理局旧址
这座欧洲新古典主义风格建筑，见证了广州邮政历史的变迁与发展。

南方大厦
南方大厦是1922年建成的中国第一座钢筋混凝土框架结构的高层建筑。

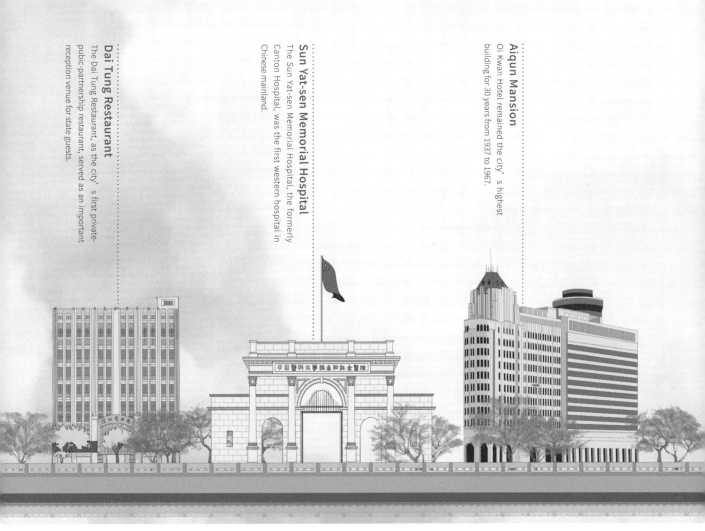

Aiqun Mansion
Oi Kwan Hotel remained the city's highest building for 30 years from 1937 to 1967.

Sun Yat-sen Memorial Hospital
The Sun Yat-sen Memorial Hospital, the formerly Canton Hospital, was the first western hospital in Chinese mainland.

Dai Tung Restaurant
The Dai Tung Restaurant, as the city's first private-pubic-partnership restaurant, served as an important reception venue for state guests.

大同酒家
大同酒家是广州第一间公私合营的食肆，是宴请各国贵宾的重要场所。

中山大学孙逸仙纪念医院
中山大学孙逸仙纪念医院的前身为博济医院，是内地第一所西医院，也是中国西医学史的开端。

爱群大厦
爱群大厦，在1937年至1967年间保持了30年『广州第一高楼』的地位。

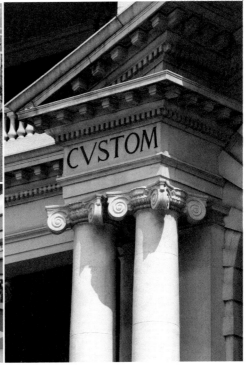

| 1 | 2/3 | 4 |

1. 在老广的记忆中，广州的清晨是从珠江边大钟楼的悠扬钟声里开始的。
2. 依水而筑的新古典主义的海关大楼。
3. 除了建筑美观，馆内的藏品丰富是广州对外贸易的重要缩影。
4. 新古典主义推崇古希腊、古罗马建筑艺术，所以"CUSTOM"的"U"，按古罗马惯例写成"V"。

Site of Former Guangdong Customs

Year Built: 1916
Add: 29, Yanjiang Xilu, Liwan District

Built in 1916, the Mansion of the Canton Customs is also nicknamed Clock Tower to the local people. It became the landmark of Changdi once completed. An excellent example of modern western-style building in Guangzhou, it is also a representative view included into the city's new Eight Scenes in the new century.

一百年前，它的门前万舸争流

粤海关旧址

中国现存最古老的新式海关大楼——粤海关大楼一落成就成为长堤的地标，因广受好评，直接影响了其后落成的上海、武汉等地的海关大楼风格。至今它仍是新世纪羊城新八景之一"珠水夜韵"的标志性景点。

年代 1916
地址 荔湾区沿江西路 29 号

1757年开始清政府仅留广州一个口岸进行对外贸易，一口通商时期，珠江上各国商船穿梭如织，十三行代理制废除后，粤海关直接接管外贸事务，其地位举足轻重。

为了照顾驻扎在沙面的西方客商，1860年粤海关迁至此处，现存的旧址大楼于1914年奠基、1916年落成。主要建材由英国进口，柚木门和地板、花岗岩立面，逾百年仍气派如初。典雅的爱奥尼巨柱，庄重的希腊式三角形山花，三层挑高的外廊空间，都可将浩渺珠水尽收眼底，空间简洁端庄，被视为广州近代西洋建筑的典范。

因塔楼为高耸的钟楼，它又被广州人亲切地称作大钟楼，珠江上航行的商船远远就能看到这座欧式新古典主义风格的大钟楼，威严自生。

岁月漫长,寄封信给时光

广东邮务管理局旧址 /

这座欧洲新古典主义风格建筑,见证了广州邮政历史的变迁与发展,是省级文物保护单位。

年代 始建于 1916 年,1942 年重建后竣工
地址 荔湾区沿江西路 43 号

| 1 | 4 | 5 |
| 2 | 3 | |

1—3.广东邮务管理局旧址,既是邮政历史的博览馆,也是滨水慢行路径上的重要景点。
4.从二、三层的外廊凭栏眺江,江风徐徐。
5.这栋新古典主义的经典之作,端庄耐看。

Site of Former Guangdong Post Administration

Year Built: 1916
Add: 43,Yanjiang Xilu, Liwan District

This European neoclassical style building witnessed the evolution of post services in Guangzhou.

广东邮务管理局旧址在长堤西岸,与西边的粤海关旧址和东侧的爱群大厦,沿珠江江岸形成高低错落的滨水天际线。

最早这里曾是大清邮政广州总部所在,火灾焚毁后,崭新的邮政大楼和粤海关旧址大楼同时竣工,它的建筑设计师曾是粤海关大楼设计师的助手,两栋比邻的典雅建筑,同样是西方新古典主义的耐看范本。

大楼在日据时期被烧毁,1939年循原貌重建,1942年竣工。大楼首层基座为黄褐色花岗岩,二、三层以爱奥尼柱贯穿,向江的南侧有前廊遮阳、防雨,柱廊为斩假石材质,在当时材料运用相当摩登。

高楼接天宇,白云映江心

南方大厦 | 大同酒家 | 塔影楼 | 爱群大厦 | 中山大学孙逸仙纪念医院

高楼林立的长堤曾是广州繁华的商业中心,这里仍完好保存着一批独具魅力的老字号和独具历史意义的老建筑。

Built in 1922, the Nanfang Building was the first high rise of reinforced concrete frame structure in China. The Dai Tung Restaurant, as the city's first private-pubic-partnership restaurant, served as an important reception venue for state guests. The Taying Building is the place Dr. Sun Yat-sen once worked at.

第一章 | 江岸
Chapter 1: Riverfront
Landscape and Attractions

南方大厦

1922年建成的南方大厦，前身是旅澳华侨蔡氏兄弟创办的大新公司。大新在香港、上海均有分号，上海大新更是后来居上，成为四大百货公司之首。落地广州，蔡氏兄弟可谓不惜血本：中国第一座钢筋混凝土框架结构的高层建筑、广东第一座使用电梯的高层建筑、当年的广州第一高度……林林总总皆令时人惊艳。

南方大厦外立面为当时盛行的水刷石饰面，风格为西方折中主义，线条雄浑、体量宏大，在很长的一段时间里，南方大厦都是繁华热闹的珠江西堤上的人头攒动风景。

西堤沿线的广场扩宽后，临江的经典建筑群落与珠江之间形成开敞的公共空间，江水映衬云影，建筑依偎大树，江风入室、行人悠然，满堤是看不完的风景。

大同酒家

大同酒家建成于1938年，1942年餐饮业翘楚冯俭生盘下此处，以其香港同名酒家重新命名、焕发新生。日本投降之后，大同酒家由冯俭生的好友——茶楼大王谭杰南接手，从此以珍馐奇馔、豪气装修扬名，一时间长堤边上的大同酒家，政要豪客云来。1955年大同酒家成为广州第一间公私合营的食肆，多款菜点入选《中国名菜谱》。

塔影楼

因倒影似塔，此楼得名"塔影楼"，塔影楼建成已百年，当年陈白沙购下联兴码头，依岸兴建此楼做办公和居住用，孙中山先生也曾入住。塔影楼共五层，其中四层为西式洋房、顶层为中式四檐滴水，式样中西合璧。沿线建筑，数它最为亲水，珠水俯首可掬，江风拂面自来。

1

2 | 3 | 4

1. 昔日长堤，最是繁华热闹之地。
2. 今天仍觉开阔大气的南方大厦。
3. 昔日贵客接踵而至的大同酒家。
4. 陈白沙故居塔影楼。

Aiqun Mansion remained the city's highest building for 30 years from 1937 to 1967. The Sun Yat-sen Memorial Hospital, the formerly Canton Hospital, was the first western hospital in Chinese mainland.

| | 2 | 3 |
|1|4|5|

1.保持了30年"第一"的爱群大厦。
2—3.立面的直线条处理，使爱群大厦更显挺拔。
4—5.孙逸仙纪念医院的主体建筑博济楼，正面和后座皆显大方典雅。

爱群大厦

爱群大厦在1967年前的30年间，一直保持着"广州第一高楼"的地位，它亦是广东最早的全钢架结构高层建筑，首届至第十届广交会开、闭幕酒会都是在这里举行的。

爱群大厦由中国建筑师陈荣枝和李炳垣设计，外立面采用连续的竖向线条、顶部成梯形的长窗，带有哥特复兴的向上挺拔之势；又因地制宜地加入骑楼和天井等岭南建筑元素，将地形的制约巧妙化解，使昂然立于江畔的爱群大厦与珠江交相辉映，犹如迎江风猎猎前行的巨轮。

中山大学孙逸仙纪念医院

中山大学孙逸仙纪念医院的前身为博济医院，是中国第一所西医院，也是中国西医学史的开端，孙中山先生曾在此学医并开展革命运动。

1935年，医院设立了"孙逸仙博士开始学医及革命运动策源地"纪念碑，同年建筑师黄玉瑜采用西式建筑手法设计了医院的主要建筑——博济楼。博济楼为西方折中主义风格，立面为水刷石，后座体形规整。正面窗户皆面江，中央门廊以巨大爱奥尼柱支撑，横向的通长挑檐的檐托密排齿饰，颇具装饰美感。

1	3
2	

1. 大元帅府曾是中国最早的一批水泥厂，现代工业在清代已然萌芽。
2. 在这条走廊上，广州被孙中山先生构思成中国第一个现代化之城。
3. 依江而建的混凝土建筑群，面朝一城繁华。

Memorial Museum of Generalissimo Sun Yat-sen's Mansion

Year Built: Qing Dynasty
Add: 18, Dongsha Jie, Fangzhi Lu, Haizhu District

Sun Yat-sen lived here for years and made many important decisions.

开门见江的传奇府邸
大元帅府旧址

旧时为广东士敏土厂办公楼，后为孙中山的大元帅府，孙先生曾在此多次颁布重要决策。

年代 清代
地址 海珠区纺织路东沙街18号

从江岸滩涂到清末中国第二大的水泥生产厂——广东士敏土厂，再到1917年、1923年两度成为大元帅府，这座百岁府邸充满传奇色彩，孙中山先生曾在此做出过许多重大决策。目前这里成为孙中山大元帅府纪念馆，是广州城市文化旅游的新热点。

2010年大元帅府前广场扩建完成后，大元帅府开门见江，中西合璧的建筑群与广场景观、珠江接连，一气呵成、气势壮阔。广场上古榕参天、江风习习，人们爱在此驻足、嬉戏，安享时光的福荫。

芳草连天碧，江波接天远
江畔学府育人已百年

中山大学南校区

这所以孙中山先生之名命名的百年学府，吸纳了很多抱持济世救国之志的人才。他们饱汲养分，怀着伟大理想、宏大抱负，最终从这里奔赴祖国各地，挥洒青春与才干。

年代 20世纪初
地址 海珠区新港路康乐村

珠江南岸的中山大学南校区，起源于广州著名的格致书院，1904年开始，书院搬到江边幽静之地康乐园，发展成岭南大学。岭南大学遵循西方近代大学的校园规划，校舍沿中轴线布列，北向直通珠江。1952年，岭南大学相关院系与国立中山大学文理科院系合并，从此康乐园校区以中山大学之名，名望远播。中山大学南校区沿承了原岭南大学的规划与建筑手法，整体按南北向的纵轴依次布列，充满庄严的仪式感。

"中大码头"原名康乐码头，孙中山先生三次乘船到岭南大学演讲，都由中大码头上岸。中大北门广场依托码头而建，车行道在广场底下通过，使得广场更完整、与水岸更亲近。古木、江水、码头、红砖校舍，此处广州城的肌理，有时光的温度，有书墨的芳香。

1	3
2	4

1. 清末兴建的马丁堂，端庄大方，是学子最爱留影之所。
2. 每当凤凰花开时节，一批批要做大事的中大人，告别母校，投身浩荡世界中。
3. 青葱校园里，曾有多少学术泰斗、共和国脊梁心耕不辍。
4. 明代的进士牌坊在广州的城市改造之际，在中大找到了庇身之所。

Sun Yat-sen University South Campus

Year Built: Early 20th Century
Add: Kangle Village, Xingang Lu, Haizhu District

On the century-old campus named after Dr. Sun Yat-Sen, the beautiful red brick academic buildings witnessed the hard studies of many ambitious young people with distinguished talents. With the determination to devote themselves to the country, they pursued their careers in different fields and made their due contributions to our country after graduation.

老空间里的新时尚

信义国际会馆 | 太古仓 | 琶醍 | 红专厂

因珠江航运便利，沿江留有大量旧时的工业遗产，为保存其独具历史意义的建筑，现在已有不少项目活化成活力十足、艺术氛围浓厚的创意园区。

信义国际会馆：荔湾区芳村大道下市直街1号，建于20世纪60年代
太古仓：海珠区革新路124号，建于20世纪初
琶醍：海珠区阅江西路磨碟沙大街118号，建于20世纪80年代
红专厂：天河区员村四横路128号，建于20世纪50年代

信义国际会馆

信义国际会馆位于珠江西部最开阔的白鹅潭畔，原是广东水利水电厂高大宽敞的苏式厂房群，现已改造为创意产业园。信义会馆是广州市首例工业遗产活化，设计既保留了建筑的精髓，又增加了时代的新元素，整个园区，建筑语言整洁明快，合抱老榕树、临江木栈桥、宽阔白鹅潭、精妙小园景，再加上老枕木铺院、旧青砖填地，时尚元素与西关人文景观融为一体，成为广州活化工业遗产的一个有益的尝试。

1. 太古仓。
2. 信义国际会馆。
3. 太古仓。

太古仓

20世纪初，英国太古洋行修建了码头及砖木结构的大体量仓库——太古仓，为其代理的轮船公司专用。收归国有之后的1955—1960年间，太古仓码头是广州内港区最繁忙、船舶到港密度最大、吞吐量最高的一个码头。2007年，太古仓成为广州一个集文化创意、展贸、休闲娱乐于一身的场所。仓立面保留原有红砖清水墙，充满工业韵味，临江绿地将码头串联成片，形成连续的滨水景观带。

New Trend in Old Spaces

Xinyi International Club: 1, Xiashi Zhijie, Fangcun Dadao, Liwan District
Pacific warehouse: 124, Gexin Lu, Haizhu District
Party Pier: 118, Modiesha Dajie, Yuejiang Xilu, Haizhu District
Redtory: 128, Yuancun Sihenglu, Tianhe District

Being revitalized industrial legacy from last century, the Xinyi International Club, the Pacific warehouse, the Party Pier and the Redtory have become dynamic creative parks or art communities that preserve unique historical buildings.

Based on the Pearl River - InBev International Beer Museum, the Party Pier is a beer-themed creative art community. The riverfronts are designed into diverse public spaces at different vertical levels.

The Redtory on the east shore of the Pearl River well preserves the architectural features of the industrial era and offers more interactive art spaces.

1	2
	3

1.坐看一江璀璨星光的琶醍。
2—3.工业历史气息浓郁的红专厂。

琶醍

琶醍是一个以啤酒文化为主题的创意艺术区，主体是华南最大的啤酒博物馆——珠江-英博国际啤酒博物馆。园区空间立体多元，有展现啤酒魅力的博物馆，有特色餐饮，还有展览空间，而景观最好的江岸沿线则还岸于民，建筑群临江部分采取退台的设计手法，最大限度地保证亲水空间的公用性，而层级递进的天际轮廓线，又增加了整体项目的趣味性。对岸则是江之南的万家灯火，坐看一江璀璨星光。

红专厂

红专厂沿用了当年中国最大的罐头厂——广州罐头厂的苏式旧厂房，厂房区的数十栋高大建筑，从20世纪50年代陆续开始建造，现在已改造成一个集设计、艺术、文化及生活于一身的创意园。红专厂南门外的珠江北岸建有亲水平台，将江景引入园内。江边设置绿地，绿地上放置工业设备改造而成的景观小品，工业历史的气息随处可见，与琶洲国际会展中心隔江相望，为珠江沿岸增添了后工业风的时尚气息。

White Swan Hotel

Year Built: 1983
Add: 1, Shamian Nanjie, Liwan District

Designed by a "dream team" comprising the Lingnan School's leading architects and engineers, this first Sino-foreign JV five-star hotel in China offers a successful example of developing tall building on sandbanks. With concise and elegant appearance, it remains a fascinating view by the riverside for decades.

从前在这里吃顿饭，可以回去炫耀很久

广州白天鹅宾馆

这个由岭南派建筑大师组构的梦之队设计的我国第一间中外合作的五星级酒店，突破了当时许多技术上的难题及观念上的藩篱，向外表达自强奋发的决心，向内则释放改革开放的信号。而自1983年开业以来，白鹅潭边的白天鹅，创造了很多个"第一"。

年代 1983
地址 荔湾区沙面南街1号

1	3	5
2	4	

1—2.珠江入羊城，白天鹅宾馆面向的白鹅潭江面最宽广，视野最为开阔，亦是从这里，内外航道分流，内航道留给两岸渡轮与观光客轮行走，繁忙的货运船只从外航道竞航。
3.在设计上，开阔的珠江美景，尽可能地引入建筑中。
4.在室内空间造出一派林木郁葱的岭南风光。
5.故乡水是最受海内外游客喜爱的拍照打卡景点。

多年以后的今天从远岸回望，将出入白天鹅宾馆的车辆与沙面岛隔阻开来的引桥，现在与江与榕、与整座岛，都和谐地融为一体；在技术和资金都匮乏的年代，筑引桥绕岛行车，填滩涂另筑高楼，做到最大限度地保留了"万国建筑博览馆"沙面岛原有的历史肌理，在高度和表皮上，也做了最大的克制，所以长期以来，白天鹅宾馆褒奖之声不竭。

白天鹅宾馆很耐看：由岭南建筑大师莫伯治和佘畯南设计——立面的水平窗内向倾斜，使简单的混凝土和铝合金窗外表呈现轻盈的韵律感；同时因地制宜地将岭南园林引入室内庭院，自然山水意境与建筑空间融为一体，中庭的"故乡水"是游人常盛不衰的打卡胜地。

白天鹅宾馆很传奇：身处南方边陲之城，前后共接待了40多个国家的100多位元首。

白天鹅宾馆很奢华：建店之初10万种物资多数进口；白天鹅宾馆也很节省：层高节缩、长腰鼓形布局，空间利用率极高。

因时代机遇，因人和自然合力，饱览浩渺碧波、坐拥老树绿荫的白天鹅，是难以逾越的典范。

乐音萦耳，江畔滔滔乐章

星海音乐厅

在珠江上静谧优雅的小岛——二沙岛上的星海音乐厅，以我国著名音乐家冼星海命名，是我国音效最好的专业音乐厅之一。除了专业的音乐演出之外，星海音乐厅亦多有普及型的音乐演出。

年代 1998
地址 越秀区二沙岛晴波路33号

从20余万平方米的绿地公园，到星海音乐厅、美术馆等等公共空间形成珠江边上的艺术聚落，沐艺术之薰，享自然之美，二沙岛散发着静谧优雅之魅力。

依江而建的星海音乐厅是广州的音乐文化地标，主体的交响乐演奏厅是中国最早的专业交响乐演奏厅之一，也是目前国内最大的纯自然声演奏厅。

星海音乐厅充满现代感的双曲抛物面几何体结构，如掀盖欲弹的三角钢琴，又如江上振翅的天鹅，雄伟壮丽、娴雅幽静。

Lingering Melody by the Pearl River

Ersha Island
Xinghai Concert Hall
Year Built: 1998
Add: 33, Qingbo Lu, Ersha Island, Yuexiu District

The Xinghai Concert Hall on the Ersha Island, a tranquil and charming isle in the Pearl River, is named after the renowned Chinese musician Xian Xinghai. As one of the best-performing professional concert halls in China, it also hosts musical performances by amateurs in addition to those by top-level musical professionals.

Guangdong Museum of Art

Year Built: 1997
Add: 38, Yanyu Lu, Ersha Island, Yuexiu District

Once the largest art museum in China and the first in the country to present the contemporary art to the general public, it was recognized as one of China's best art museums at that time. It has committed itself to the public art education and has been working persistently in improving the art literacy of the general public.

这座美好的庭院，是中国当代艺术史上响亮的名字

广东美术馆 /

它曾是全国最大型的美术馆，也是内地最早向民众推送当代艺术的美术馆，被公认为当时国内最好的美术馆之一。广东美术馆也非常注重公共教育，坚持不懈地普及美育工作。

年代 1997
地址 越秀区二沙岛烟雨路 38 号

广东美术馆面朝珠江，与星海音乐厅、广东华侨博物馆等等面向公众的公共空间相邻，在二沙岛形成一个文化氛围浓厚的艺术带。美术馆总体建筑面积2万多平方米，是中国面积最大的美术馆之一，其旨在呈现中国当代艺术的广州三年展颇具国际影响力，同时馆方在把艺术推向更广的公共空间上，也做了诸多有益的尝试。

这座最早入选4A景区的美术馆，是以庭院体系为美学表现的建筑群体，立面色彩呈灰绿色，与相依相傍的珠江相呼应。从岛中央的公园绿地，到一步之遥的珠江江堤，形成了广阔无边际的展览平台和休憩平台。人们在领受美学熏陶的同时，亦得以饱览珠江美景，悠享习习江风。

1. 从馆外的大广场，与珠江江岸相接。
2. 室外园林空间与室内展览互为补充，整个空间充满律动。

Chapter 1: Riverfront
Landscape and Attractions

Guangdong Museum of Chinese Nationals Residing Abroad

Year Built: 2002
Add: 32, Yanyu Lu, Ersha Island, Yuexiu District

The robust development of Guangdong would not be possible without the great dedication from the over 30 million Cantonese residing all over the globe. This museum is built in memory of their remarkable dedication and serves as a living example to depict the Cantonese Spirit together with the art museum and concert hall on the island.

广东侨民，
一部奋斗与奉献的史诗

广东华侨博物馆

广东的蓬勃发展，离不开3000多万分布在世界各地的粤籍华侨的无私奉献，这座纪念华侨们卓越奉献的博物馆，和毗邻的美术馆和音乐厅一起，成为解读广东精神的文化地标。

年代 2002年建成，2011年正式开馆
地址 越秀区二沙岛烟雨路32号

广东是中国对外交往的门户，是中国侨民最多的省份，"有海水的地方就有粤籍华侨"，从广东华侨博物馆内的粤籍华侨分布地图上，可以看到粤籍华侨以顽强的生命力，在地球的各个角落落地生根。近代以来，广东的发展离不开3000多万粤籍华侨的帮助和奉献。为了弘扬侨胞精神，加强人民与侨胞的桑梓情谊，广东华侨博物馆经筹备多年后向公众开放，它是目前国内唯一一所省级的专业华侨博物馆。

博物馆的建筑造型方圆结合，寓意天圆地方、四海一家。馆内文物翔实，走进正门，一艘帆船首先映入眼帘。从博物馆步道，可直通珠江，形成景观视廊，当年勤劳勇敢的粤籍侨民便是这样乘远洋轮船借江通海，在世界各地奉献自己的聪明才智和青春年华，向世界传达一个自强不息、大写的中国。

1. 馆内的珍贵资料，向后人讲述了老一辈华侨的血泪奋斗史与拳拳爱国心。
2. 当年，多少广东华侨乘坐远洋轮船从珠江出发，开启自己的寻梦之旅。

今天的广州风光片，
它们总是主角担当
Canton Tower / Haixinsha Island
广州塔｜海心沙

第一章 | 江岸
Chapter 1:Riverfront
Landscape and Attractions

1	3
2	

1—2.沿着珠江构图，600米的广州塔是主角的不二选。
3.海心沙便如珠江里的巨轮迎着东方旭日扬帆。

Canton Tower and Haixinsha Island

Year Built: 2009
Add: 222,Yuejiang Xilu, Haizhu District

Being China's No. 1 tower and the world's No. 2 in height, the 600-meter Canton Tower holds many world records. For example, it has the world's highest and longest sky walk steps, the world's highest revolving restaurant, ferris wheel, vertical speed drop and outdoor observation deck.

Haixinsha, an isle in the Pearl River, is the most popular civic park in Guangzhou and the first-choice venue for major public events.

今天的广州风光片，
它们总是主角担当

广州塔 | 海心沙

总高度达600米的中国第一高塔——广州塔，拥有多项世界纪录：世界最高最长的空中漫步云梯，世界最高的旋转餐厅，世界最高的摩天轮、垂直速降游乐项目和户外观景平台，提升高度最高的电梯。

海心沙则与广州市最大的广场——花城广场，连接成片，许多重要的大型演出都在这里举行。

地址　广州塔：海珠区阅江西路222号，2009年建成
　　　海心沙：珠江内江心沙洲

矗立于新城市中轴线与珠江景观带交会点的广州塔，是今天珠江边上最为人熟知的明星。

除了各种世界纪录，广州塔的施工难度之大，亦开创了许多纪录：

塔身钢结构外框筒的24根钢立柱向上逆时针转动45°、延伸454米时，底部直径2米缓变为顶端直径1.2米，而其误差控制在5毫米内；

钢结构每一截面的三维数据都在变化，一万多构件无一相同，加工、制作、施工难度可想而知；

高强度高性能的混凝土，被泵送到450米，开创国内混凝土泵送最高的新纪录；

绚丽多变的外壳是高强度钢化夹胶玻璃，为保证其线条的流畅，每一处玻璃尺寸都随曲线变化而变化；

混凝土的核心筒与钢结构的外框筒，因材料特性的差异，势必造成沉降不一致，高度难以统一；

内骨架由5万吨厚钢板焊接而成……攻克所有难题的广州塔，亦成为21世纪建筑史上的里程碑。

海心沙这处珠江内的沙洲，是新晋的广州旅游胜地，珠江里的这颗明珠，曾在亚运会举行开闭幕式时惊艳登场。亚运会后的海心沙，已经成为广州人最为喜爱的市民公园，身处其中，人们可零距离欣赏美轮美奂的江景。夜幕降临，璀璨的灯光倒映在水中，成珠江水中最迷人的星光。

第一章│江岸
Chapter 1:Riverfront
Landscape and Attractions

珠江里宁静的砾石，
都市里流动的歌

广州大剧院 /

由第一位获得"普利兹克建筑奖"的女性建筑师设计的广州大剧院，被誉为"世界十大歌剧院"。

年代 2010
地址 天河区珠江西路1号

广州大剧院是著名建筑师扎哈·哈迪德在中国的第一个中标实施方案项目，一举中标不到半年时间，世界建筑设计的最高荣誉"普利兹克建筑奖"，颁给了这位从阿拉伯世界走出来的天才建筑师，她也是自1979年奖项设置以来第一位获此殊荣的女性。

扎哈的方案最打动广州人的地方，就是强调了大剧院与珠江的关系：大剧院宛如久经珠水冲刷的灵石，扎哈所擅用的非线性、流动性的设计形态，正与珠江流域冲积平原的肌理非常契合，又与大剧院所在的地理位置完美呼应。在理念上因循中国传统"师法自然"的思路，又以充满异型设计的钢结构、现代感十足的玻璃和石材立面，无论晨昏与珠江交相辉映，使得这一组"珠水砾石"的表达，产生了深深的感染力。室内的声学效果设计是声学大师哈罗德·马歇尔的佳作，该音效获得全球建筑界及登台艺术家的极高评价。

Guangzhou Opera House
Year Built: 2010
Add: 1, Zhujiang Xilu, Tianhe District

Designed by the Iraqi-born British Architect Zaha Hadid, the first female winner of the Pritzker Architecture Prize, the stone-and-glass-clad complex is reminiscent of two rocks washed away by the water in the Pearl River. It fits well into the urban context and has become one of the most favorite shooting scenes of the photographers both in the day and at night.

滨水新岸双生塔

广州国际金融中心 | 广州周大福金融中心 /

广州人将珠江新城濒临珠江的两栋摩天大楼——广州周大福金融中心和广州国际金融中心，亲切地简称为东塔和西塔，两塔与珠江对岸的广州塔，这三个珠江边超400米高耸入云的新地标，被称为广州新三塔。

年代 2010 年 | 2014 年
地址 广州国际金融中心：天河区珠江西路 5 号
　　　 广州周大福金融中心：天河区珠江东路 6 号

广州国际金融中心（西塔），高432米，广州周大福金融中心（东塔），高530米，双塔沿珠江新城中轴线对称分布，中间以中轴线上的巨大花园广场——花城广场相隔。

西塔在超高层建筑中首次应用具有良好的抗侧刚度和抗震性能的钢管混凝土巨型斜交网格筒中筒结构体系，主塔楼的全隐框玻璃幕墙面积达8.5万平方米，使西塔外立面精美流畅，典雅现代，随着天光及灯光的变化，呈现出水晶般的瑰丽晶莹。通透的立面也将珠水美景引入建筑，每一扇窗都是取景框，描绘着不同角度的江景。

东塔的外立面则采用玻璃幕墙加石材、螺纹陶瓦板，既有效减少太阳光的反射，螺纹陶瓦板又更能适应广州多雨、潮湿的气候，对环境亦更加友好地呼应。它还采取"之"字形的退台设计在不同楼层形成空中花园，大厦顶部天台开敞，可从各面远望、饱览珠江两岸风光。

1	3
2	

1—2.无论阴晴、无论晨昏，珠江两岸的东西二塔和广州塔这三座地标新三塔，巍峨参天、瑰丽夺目。
3.新中轴线上的东西两座摩天楼，除了高颜值，亦开多项业界先河。

Twin Towers by the Pearl River

Year Built: 2010 | 2014
Add:
 Guangzhou International Finance Center: 5, Zhujiang Xilu, Tianhe District
CTF Finance Center: 6, Zhujiang Donglu, Tianhe District

The CTF Finance Center and Guangzhou International Finance Center, known to the local people as the East Tower and the West Tower respectively, form two of the new lofty towers in Guangzhou and the city's new landmarks together with the Canton Tower rising on the opposite side of the Pearl River.

The two modern buildings in striking heights employ numerous high technologies. The West Tower once ranked No. 1 among the individual high-rises in terms of the area of the hidden-frame façade, while the East Tower perfectly brings the view of the Pearl River into the interior via setback design.

第一章 | 江岸
Chapter 1 Riverfront
Landscape and Attractions

珠江边上的记忆宝盒

广东省博物馆

　　广东省博物馆是一座省级的综合性博物馆，是广州的标志性文化设施之一。

年代 2010
地址 天河区珠江东路 2 号

　　广东省博物馆倚靠珠江，与西侧的广州大剧院、北侧的广州图书馆等等公用建筑，以及新城市中轴上的花城广场、临江的绿化带一道，构成一片依江营造、景观无敌的文化艺术走廊。

Guangdong Museum

Year Built: 2010
Add: 2, Zhujiang Donglu, Tianhe District

Inspired by the ivory ball, a traditional Cantonese craftwork, the interior and exterior structure of the museum resemble those of an ingenuously devised and hollowed-out ivory ball. With changing light and shadow effect on the building skin and the interconnected interior spaces, it also resembles a treasure box by the Pearl River, preserving Guangdong Province's top cultural and natural heritage.

博物馆的建筑设计以广东传统的工艺品象牙球为灵感，整个内外结构如同剔透的镂空象牙宝盒。

这个宝盒，安身于从花城广场延伸过来的草坡台座中，与珠江相呼应，它收纳着广东地区最高级别的文化遗产和自然遗产，陈列展览以广东历史文化、艺术、自然为三大主要陈列方向，是了解广东历史文化背景的重要窗口。

博物馆为应对珠三角潮湿的气候，首层架空，建筑外表用色为岭南民居常用的传统青灰色，减少了向四周环境的太阳热能反射，窗顶和窗侧不规则的深凹开窗，实现遮阳和采光的有机统一。内外光影有趣呼应，内部空间层层相扣，层次深远，层层迂回、递进，吸引着观众层层前进。

宝盒外的广场延至珠江边，风自珠江来，吹拂游人面。而无论晨昏，宝盒映衬江面皆成诗画。

1.广州人把寸土寸金的珠江新城最为珍稀的临江地段，留给了这个汇集南粤大地珍宝的宝盒。
2.这座储存着广东集体回忆的博物馆，是新规划的珠江景观线上，最美的风景。

第一章 | 江岸
Chapter 1: Riverfront
Landscape and Attractions

CBD 的大手笔，
献给现代人通往理想的阶梯

广州图书馆

广州图书馆是世界上规模最大的城市图书馆之一，是全国一级图书馆。同时在多元文化和本土人文等的各类特色服务上，都有极具开拓精神的尝试。

年代 2012
地址 天河区珠江东路 4 号

依水而筑的广州图书馆坐落于花城广场，处于城市新中轴线和古老珠江交会处。它与周边的广东省博物馆、广州大剧院、广州市第二少年宫形成一个景观绝伦的滨水文化区，CBD（中央商务区）江岸的景观线，被最大限度地公共化。

图书馆以"美丽书籍"为理念设计，外观如一本打开的书本，内部设计加入了广东常见的骑楼元素，使得采光更优化。整体建筑采取东西走向、独特的"之"字形优雅造型，突出层叠的建筑肌理，寓意书籍的重叠和历史文化的沉积。

开放之初，就以广州一贯的包容姿态，建立多元文化馆、广州人文馆、语言学习馆三大主题馆，增进公众对不同文化的了解、尊重。

Guangzhou Library

Year Built: 2012
Add: 4, Zhujiang Donglu , Tianhe District

One of the world's largest urban libraries and a national Class I library in China, Guangzhou Library resembles an opened book in appearance. The interior design references the commonly found local arcades to optimize the daylight.

第二章：海丝
Chapter 2: Maritime Silk Road

广州建城2200余年以来，向来是中国海上贸易的重要窗口，旧日海上丝绸之路的线索清晰可辨：莲花塔、琶洲塔、赤岗塔三座航塔安好，黄埔古港、光塔、沙面等商贸往来重镇仍在，如今，新时期的海上丝绸之路——"一带一路"上的传奇，正在续写新篇。

Since its founding over 2,000 years ago, Guangzhou has been an important window in China's history of maritime trade. The heritage of the Maritime Silk Road left from the old days can still be found in the city. The three navigation towers, namely the Lotus Pagoda, the Pazhou Pagoda and the Chigang Pagoda, are kept intact, while the major trade ports like the Ancient Huangpu Port, the Minaret (Lighthouse) of the Huaisheng Mosque and Shamian are still well-preserved. Nowadays, the Belt and Road Initiative as a new Maritime Silk Road in the new era is about to write a new chapter in history.

Nansha Tianhou Temple

Year Built: 1994
Add: 88, Tianhou Lu, Nansha District

Facing the vast Lingding Ocean, the Tianhou Temple and the Dajiao Mt. Fortress were the best message for the merchant ships travelling along the Maritime Silk Road, as they signaled their arrival at Canton, a place full of opportunities and hopes. The Fortress is also the first one that defended the navigation channels of Guangzhou.

2
1	3

1. 从大角山远眺，便是伶仃洋浩渺烟波，是联结东西的海上丝路。
2. 这条繁荣兴旺往来的贸易之道，天后的照拂是远洋人的心灵慰藉。
3. 这珠江口的门户之地，拥抱过希望，也经历过炮火。

第一眼，人们看到了祝福和希望

南沙天后宫

由海岛变桑田，这个东南亚最大的妈祖庙体现了对海洋文明的崇拜和敬畏。

年代 明代始建，1994年重建
地址 南沙区天后路88号

南沙天后宫位于南沙大角山东南麓，面朝浩瀚如烟的伶仃洋，其前身是明朝天妃庙，后于战争中被炸毁，如今的南沙天后宫是1994年仿照清代宫殿式建筑形制建造的。

先秦时期，南沙一带水系发达，如今南沙地势较高的小山岗是以前的海岛。当时的人民以农耕和渔猎满足生存需求，形成了对海神天后的崇拜与信仰。

南沙天后宫被誉为"天下天后第一宫"，是东南亚最大的妈祖庙。建筑倚山营造，对称布局，顺山势而筑，错落有致。天后宫广场上矗立的天后石雕圣像由365块花岗岩石砌成，高达14.5米，象征天后一年365天保佑风调雨顺、国泰民安。

Three Ming Pagodas as Navigation Marks on the Maritime Silk Road

Year Built:
From Ming to Qing Dynasties
Add:
Pazhou Pagoda: Xingang Donglu, Haizhu District
Chigang Pagoda: Yiyuan Lu, Haizhu District
Lotus Pagoda: the Lotus Mt., Panyu District

The Fengshui pattern known to the locals as "Three pagodas as three fortresses to lock the Pearl River" came into being in the Ming Dynasty. Three pagodas were built along the Pearl River at that time, for the local people, as the loyal Fengshui believers, believed that the three pagodas could stabilize the estuary and ensure the success of the local people in both imperial examination and business. In fact, the three pagodas also functioned as navigation towers for the merchant ships along the Maritime Silk Road.

1—2.明万历年代兴建的琶洲塔，立于琶洲岛的小山岗上，居高镇守珠江。
3.八角形楼阁式砖塔，工艺细腻比例秀美。
4.塔基的托塔力士，神态威武。

明代的三座古塔，曾是海上丝绸之路的导航标

琶洲塔｜赤岗塔｜莲花塔

400年来，三座古塔锁江镇海，见证了广州海上贸易的繁荣与变迁。

年代 明—清
地址 琶洲塔：海珠区琶洲新港东路
赤岗塔：海珠区艺苑路
莲花塔：番禺区莲花山

从汉代至清代，广州得天独厚成为海上丝绸之路重镇，到了清朝甚至一口通商、独领风骚。400年前，当各种肤色的外国商人乘着商船从外洋穿过虎门大关、进入珠江航道，透过西洋望远镜望见这三座塔，他们就知道：Canton到了。

这三座塔就是莲花、琶洲和赤岗三塔，也是昔日广州诗云"白云越秀翠城邑，三塔三关锁珠江"中所指的三塔。

清代羊城八景之一的"琶洲砥柱"，正是珠江边琶洲上的琶洲塔。琶洲原是珠江中的小岛，因山形似琵琶而得名，琶洲塔建于明代，和赤岗塔同为八角形楼阁式砖塔。相传水中常有金鳌浮出，所以此塔原称"海鳌塔"。琶洲岛上的琶洲塔，为海上丝绸之路上商船指路引航。

400年沧海桑田，如今的琶洲集会展经济、创意产业、新技术产业与观光旅游于一身，海上丝绸之路再添迷人风姿。

1—2.明代地标赤岗塔与600米的新地标广州塔相邻,历史的记忆,沿江绵延。
3.雄踞珠江口的第一塔——莲花塔。

The three Ming pagodas rising at the once barren suburbs are now embraced by the bustling city life. Standing at the top of hills, they still offer fascinating views of the Pearl River even today.

　　莲花塔与琶洲塔、赤岗塔三座风水宝塔沿珠江鼎足而立,犹如航船三根桅杆,俗名"三枝樯"。建于明万历的莲花塔,八角楼阁式、粉墙红柱、绿琉璃瓦、八角攒尖顶,因建于莲花山主峰上,雄踞珠江入口,有"省会华表"之称,从外洋来广州,望见莲花塔,就知道广州在望。

　　楼阁式的青砖塔赤岗塔建于明万历年间,高约50米,平面为八角形,外观9层,塔内17层。基座八角用红砂岩垒砌,均镶有西方人形象的托塔力士,正是明代对外交往的佐证。

　　走在人流兴旺、商业气氛浓厚的赤岗,不知多少人留意到这座拥有400年历史的明代砖塔?很多人重新认识到这座塔也是因为距它北面约900米处的小蛮腰(广州塔)吧?它们一古一新比邻而立于城市新中轴线南端,代表着传统与创新的广州。

South Sea God Temple

Year built:
From Sui to Qing Dynasties
Add: 22, Xuri Jie, Miaotou Village, Huangpu District

The temple was built in the Sui Dynasty to pray for a calm sea. It is the largest and earliest sea god temple so far preserved among the four ancient sea god temples in China. The Sunbath Pavilion nearby was acclaimed as one of the city's Eight Scenes in the Song, Yuan and Qing Dynasties.

1	3
2	4

1.南海神庙虽历代均有翻修，但历代的文物珍藏依旧相当可观。
2.从南海神庙献了香火，至海不扬波牌坊出来，就是狮子洋，进则入城，出则入海。
3.从半空看南海神庙，可以体会到当年的壮阔宏伟。
4.南海神庙的波罗诞至今仍是受乡民喜爱而至的隆重节庆。

旧日海上商贸的保护神

南海神庙

隋朝时，建南海神庙以镇南海，现为中国古代四大海神庙中唯一留存下来的海神庙。

年代 隋—清
地址 黄埔区庙头村旭日街 22 号

南海神庙又称波罗庙，濒临古代广州外港，其地理位置被韩愈称为"扶胥之口，黄木之湾"，是海上丝绸之路的始发地。

唐代以前，珠江江面广阔，隋开皇十四年（594），南海神庙应文帝诏令滨水而建，以镇南海。自明朝始，民间每年农历二月十三日会在南海神庙举行"波罗诞"，祈求海不扬波、出入平安。

南海神庙是中国古代四大海神庙中唯一留存下来最大、最古老的海神庙，其附近的浴日亭在宋元清时期被列为羊城八景之一——"扶胥浴日"。

1. 因其珠水汇流、江海交融的地理位置，黄埔古港在很长的时间里，是广州对外贸易里程中最重要的门户。
2. 水通财通，商行林立、货仓遍布的黄埔古村，也盛产读书人家。
3. 秀美的灰塑华拱。
4. 祠堂街附近的建筑有序布列，保存完好。
5. 珠水南退，但黄埔古村人，仍长于渔业。

千帆过尽，古港仍在

黄埔古港

在闭关锁国的清朝，粤海关是唯一一个对外开放的海关，而粤海关的挂号口便设于黄埔古港。

年代 清代
地址 海珠区石基村

黄埔古港位于广州市海珠区，濒临珠江。乾隆年间，清政府闭关锁国，独保留粤海关，而粤海关挂号口便设于黄埔古港，在此之后，黄埔古港便一枝独秀盛开了80年。

黄埔古港见证了海上丝绸之路的繁荣，1769年，英国人威廉·希克曾感叹："珠江上船舶运行穿梭的情景，就像伦敦桥下的泰晤士河。"

黄埔古港与水息息相关，黄埔古港依水而生，黄埔古村因水而兴。街巷以青石铺砌，配以景观小品，古朴大方，与环境完美协调。

Ancient Port of Huangpu

Year Built: Qing Dynasty
Add: Shiji Village, Haizhu District

The Ancient Huangpu Port at the convergence of the Pearl River tributaries and estuary was the city's most important gateway in China's foreign trade history. During the Canton System period, the registration office of the Guangdong Customs was located here as the one and the only for 80 years.

一口通商十三行

十三行博物馆

十三行在"一口通商"独揽中国外贸85年之久,是清朝时期东西方文化交流的门户。

年代 2016
地址 荔湾区西堤二马路 37 号广州文化公园内

十三行地处广州西关沿江地带,是自古以来的海上丝绸之路的起点,也是清廷开设海关之后,清朝商人代表政府去与外商进行交易、代收关税的商行。

作为中西方文化技术交流的开放门户,十三行存活了150多年,并在"一口通商"的时期,独揽中国外贸85年之久。

十三行博物馆在十三行的遗址(今文化公园)上建成,那些被历史湮没的耀眼辉煌,又重新出现在大众眼前。走进十三行博物馆,就走进了中西文化交融的历史时空,一件件美轮美奂的广彩瓷器、珐琅器、象牙器、玻璃画、通草画、外销扇向世人讲述了清代一口通商的粤海关盛景。

Guangzhou Thirteen Hongs Museum

Year Built: 2016
Add: Inside Guangzhou Culture Park, 37, Xidi Ermalu, Liwan District

The Thirteen Hongs at the waterfronts of Xiguan area were actually commercial houses where the Qing merchants traded with and collected the customs duties from the foreign merchants on behalf of the Qing government after the latter set up the customs in Guangdong. They were authorized to conduct foreign trade on behalf of china for 85 years due to the implementation of the Canton System.

1	3
2	4

1. 360°环形通草画,将当年十三行万舸竞渡盛况重现眼前。
2. 符合彼时西方审美的巴洛克装饰风,和清中晚时期的富丽之风结合在一起,组成了各式华美的艺术品。
3. 以十三行为主角的外销画,漂洋过海,向西方世界展现广州的繁华。
4. 不惜工本、不计物力的外销产品,至今仍是收藏市场的宠儿。

十三行烧了，但沙面还在

广州沙面建筑群/

昔日沙面的血泪史料今已寥寥，唯独岛上樟榕老树年年葳蕤。

年代 清代—20 世纪 30 年代
地址 荔湾区沙面大街

Historical Architecture in Shamian

Year Built: From Qing Dynasty to 1930's
Add: Shamian Dajie, Liwan District

The classic western-style building clusters in Shamian were one of the first national priority protected sites in China. Shamian, an alluvial sandbank by the Pearl River, became an isolated isle in the River when a canal was built to separate it from the downtown, which happened after the Thirteen Hongs were burnt down by fire.

1
―
2

1.百余岁的樟与榕，映衬着沙面的建筑，处处烙着光阴的痕、无穷的绿。
2.西江和北江在三水汇合后，蜿蜒流进广州，再从沙面对出的白鹅潭向东、向南分流向大海奔去。

| 1 | 3 | 5 |
| 2 | 4 | |

1.台湾银行旧址。
2.外廊式建筑除了美观，也很适合日照强烈的广州。
3.露德天主教圣母堂。
4.沙面外事博物馆。
5.这被珠江环抱的小岛，处处是离尘的静谧之美。

　　沙面古称拾翠洲，是由珠江冲积而成的沙洲。1861年，沙面被划为英法租界，并被凿渠成为珠江里的孤岛，与城区以河涌相隔。

　　沙面建筑风格多样，有红砖尖顶的浪漫主义建筑——海关洋员华员俱乐部、宏伟严谨的新古典主义建筑——东方汇理银行、尖拱高耸的哥特式建筑——露德天主教圣母堂、自由立体的现代建筑——白天鹅宾馆等。沙面本地居民把沙面大街装扮成一条精致的景观主轴，以古朴的铺装、雕塑、街头游园引导游人在岛内徜徉。沙面公园与沙面二街把滨水景观引入岛内，让人流连忘返。

　　1996年，"广州沙面建筑群"成为首批进入"全国重点文物保护单位"的西方古典式建筑群。

The isle is home to many elegant western-style buildings and the century-old camphor and banyan trees in lush green.

第二章 | 海丝
Chapter 2:Maritime Silk Road

重新出发，抵达另一片海洋

琶洲国际会展中心 /

亚洲最大，世界第三的会展中心。

年代 2002
地址 海珠区阅江中路 380 号

琶洲国际会展中心位于赤岗琶洲岛，创下两项世界第一：单体展馆面积最大，达39.5万平方米；钢桁架跨度最大，每个展厅的顶部由6个长达 126.6米的大跨度预应力张弦梁钢管桁架支撑着。琶洲国际会展中心是目前亚洲最大、世界第三大的会展中心。

Pazhou International Convention Exhibition Center

Year Built: 2002
Add: 380, Yuejiang Zhonglu, Haizhu District

The spectacular international exhibition and convention complex by the Pazhou Pagoda and the Chigang Pagoda, the two navigation towers on the Maritime Silk Road, created two world records for being the world's largest singular exhibition and convention facility and having the world's largest span of the steel structure. Currently, it remains the largest exhibition and convention center in Asia and the third largest in the world.

第三章：湖光
Chapter 3:Lake Parks

广州人的水畔生活仓廪丰足、恬适安康，自宋代起，人们开始评选羊城八景，生活，便花红柳绿地诗意起来。那些分布在城市各处、翡翠一样美丽的湖，既为城市增添令身心愉悦的绿和蓝，亦在功能上排内涝，调节小气候、改造大环境，为广州注入灵动的潋滟湖光。

These picturesque lakes around the city add pleasant green and blue tones to the city and function positively in improving the microclimate and urban environment for a more dynamic and inspiring urban experiences.

一城的水秀灵动，
由人工造就

湖泊——水秀灵动

除了天然的湖，广州很早开挖人工水利。1958年，广州新筑了3个人工湖（流花湖、东山湖、荔湾湖）。加上之前开辟的麓湖，形成四大人工湖。广州的人造湖是融功能和景观为一体的综合性工程，既能蓄洪排涝又创建城市景点。

Artificial Lakes Make a Dynamic and Lively Urban Experiences

Functioning for both flood control and landscape purpose, the artificial lakes in Guangzhou are indispensable attractions that make the city more attractive and livable.

| 1 |
| 2 |

1. 经数十年人力勤勉经营，东山湖往东，是繁华的新城，也是林木连绵、天水一色的宜居之地。
2. 从麓湖聚芳园远眺广州，是无尽的绿与蓝。

青山入城水相接
Luhu Park
麓湖公园

第三章 | 湖光
Chapter 3:Lake Parks

青山入城水相接

麓湖公园/

一山环秀水，半岭隐涛声。一座环湖而建的综合性公园。

年代 1958
地址 白云山风景区南麓

麓湖公园是一个环湖而建，以自然水景为特色的综合性公园。

麓湖公园充分利用连绵起伏的山势进行造景，并在山体制高点设鸿鹄楼，统领全园。山景与水景相得益彰，无论漫步在滨水木栈道还是环湖林荫道，游人都能感受到流水野趣，流连忘返。

园内有为纪念我国伟大的无产阶级音乐家冼星海而建的星海园，内有纪念馆、巨型石雕像等。

Luhu Park

Year Built: 1958
Add: South of Baiyun Mt. Scenic Area

At the southern foot of the Baiyun Mt. that extends into the city, the Luhu Lake was excavated to control the flood from the mountains and connect with the Pearl River via the Donghaochong Canal. The park has a picturesque view with flowering trees around the lake all year around.

1	3
2	4

1.借白云山余脉造景，麓湖公园有山有湖，从鸿鹄楼眺望，稠密楼宇被青山紧抱。
2—3.为泄山洪而挖建的麓湖，早已林密花繁，山、水、亭、桥、林，无一不美。
4.数十年间，麓湖公园经历次改造、升级，越发完善。

101

Chapter 3: Lake Parks

城的西边，有座飞花的湖

流花湖公园/

被誉为"鹭鸟天堂"和"岭南盆景之家"的大型公园。

年代 1958
地址 越秀区东风西路 100 号

| 1 | 3 | 1.城西的流花湖，拥流花桥和芝兰湖的旧痕故典，亦是近代中国的造园典范。
| 2 | | 2.满溢热带风情的葵堤和红桥，名列广州历史建筑名录，很有代表性。
| | | 3.流花湖，果真是花的湖。

Liuhua Lake Park

Year Built: 1958
Add: 100, Dongfeng Xilu, Yuexiu District

The Liuhua Lake is named after the Liuhua Bridge, a Nanhan attraction. It was the former Zhilan Lake which was backfilled and diminished in ancient time. After the founding of the PRC in 1949, the government called on the people to dredge and deepen the lake and build the park for sightseeing, recreation and leisure.

流花湖因南汉古迹流花桥而得名，原为古时淤塞湮没的芝兰湖，新中国成立后，政府发动全市人民疏浚积涝，建成融游览、娱乐、休憩为一体的综合性大型公园。

流花湖遵循地势辟有浮丘、芙蓉洲，并广植棕榈、榕属植物、开花灌木等，形成热带特色风光。四季鲜花争艳，湖光花色皆成画，洋紫荆、荷花、葱兰、紫薇、木芙蓉次第盛放，入冬则池杉、落羽杉转红，如霞光映水。湖光花色，不负流花湖处处飞花之美名。

著名的葵堤四季常青，湖中岛飞鸟蹁跹，西苑盆景满院，闹市中的这处绿洲，湖光流花看不厌、鸟鸣欢声听不足。

这些湖长着长着，
成了绿宝石蓝宝石

东山湖公园

东山湖公园中具有标志性的建筑，砖红的桥面迂回、曲折，可带人领略不同角度的风景。

年代 1958
地址 越秀区东湖路 123 号

Dongshan Lake Park

Year Built: 1958
Add: 123, Donghu Lu, Yuexiu District

东山湖公园在1958年曾是一片洼地和菜田，广州解放后才改造成一个文化休闲公园。

公园以湖为主，园内山水相依，湖水清澈见底，湖中有5个半岛和1个孤岛，岛间以桥相连，再与各具特色的亭、廊、榭、阁形成一体。公园中有青麻石构成的五拱桥、精致雅韵的一拱桥、落虹桥和亲水的贴水桥。九曲桥是东山湖公园中具有标志性的建筑，砖红的桥面迂回、曲折，可带人领略不同角度的风景。

1963年，因其四季如春的景色，东山湖公园被评为"羊城八景"之一——"东湖春晓"。

The park was once one of the Eight Scenes of Guangzhou for its charming view in the early spring.

荔香飘送，藏龙于湖

荔湾湖公园 /

荔湾湖公园与荔枝湾涌相连，是台、榭、廊、轩掩映于无穷青碧色的典型岭南园林。

年代 1958
地址 荔湾区荔枝湾路

Lychee Lake Park

Year Built: 1958
Add: Lizhiwan Lu, Liwan District

　　荔湾湖公园与荔枝湾涌相连，或者说，如今的新荔枝湾涌正是以荔湾湖为主体而改造的。小翠湖、玉翠湖、如意湖、五秀湖，以及湖边风情万种的荔枝树，使荔湾湖公园碧波荡漾，荔香处处；景观上八亭八桥交相辉映，路堤及小岛数量繁多、姿态各异，台、榭、廊、轩掩映于无穷青碧之中，是岭南园林的典范。

　　荔湾湖内以堤分隔水面，迷离曲折，涌畔、湖边均建有亲水平台，游人可近距离欣赏静谧水景。茵茵绿草点缀在步道两侧，野趣盎然。有心人更可找到藏龙塘的所在地，端午将至时，看"壮丁"们跃入湖中"起龙船"，祈福过后再到荔湾大戏台看一台好戏，便是老广心驰神往的盛夏礼遇。

Connected to the Lizhiwan Canal, the park has a typical Lingnan garden where terraces, pavilions, cloisters and verandas intertwine with the lush green.

第三章│湖光
Chapter 3:Lake Parks

与繁华都市照面的国家级湿地公园
海珠国家湿地公园

一座被誉为全国特大城市中心区最大的迷人的国家湿地公园。

年代 2012
地址 海珠区新滘中路 168 号

不是每个城市都能包容一个湿地公园在市中心，而海珠国家湿地公园偏偏卧在广州城市新中轴南端，从黄埔涌延伸到珠江后航道，涵盖万亩果园、海珠湖和数十条河涌，被誉为广州"南肾"，与"北肺"白云山一同构筑城市的生态屏障。

海珠湿地水网交织，是珠三角河涌湿地、城市内湖湿地与半自然果林镶嵌交混的复合湿地生态系统，是候鸟迁徙的重要通道，真正的"小鸟天堂"。园内百果飘香，绿树婆娑，岭南水乡与湿地相融，人们因地制宜，于是有了新建的特色镬耳屋与亲水平台，游人可驻可亲，观鸟赏花，悠然自得。

Haizhu National Wetland Park

Year Built: 2012
Add: 168, Xinjiao Zhonglu, Haizhu District

The park is reputed as the largest national wetland park with charming views at the center of a mega city.

在城的北边，更多的水之乐园在建设
白云湖公园 | 南湖游乐园

城市的轮廓在扩大，人口日渐稠密，昔日的城北，也有更多的水之景观，改善着人们的生活环境。

年代 2006 年 | 2012 年
地址 白云湖公园：白云区石井大道
　　　南湖游乐园：白云区广州大道北 983 号

More Water Parks under Construction in Northern Guangzhou

Add:
Baiyun Lake Park: Shijing Dadao, Baiyun District, 2006
Nanhu Lake Amusement Park: 983, Guangzhou Dadao North, Baiyun District, 2012

Along with the urban expansion and increased population density, the once marginalized northern part of the city has become home to more and more water parks which will offer better living environment to the residents.

1. 白云湖公园。
2. 南湖游乐园。

　　一路向北，满眼碧水并没停歇。白云湖公园是目前广州最大的人工湖，湖区面积超过市内荔湾湖、流花湖、东山湖、麓湖面积的总和。白云湖水、陆域各半，是广州市北部水系建设珠江西航道引水首期工程，兼具雨洪调蓄和休闲景观的功能，是融水安全、水生态、水文化、水景观为一体的综合性水利工程。

　　白云湖独具岭南特色，堪称广州"西湖"。湖边观赏生态林超过百亩，26个岛屿组成的"岛链"在湖面星罗棋布。京广铁路跨湖而过，围起150亩人工湿地，绿色生态尽收眼底。

　　同拥"白云"的，还有南湖。旧称磨刀坑水库，依着白云山西麓，是广州市重要的小型水库。环水库建有南湖游乐园，留存着一代广州人的美好童年回忆。当年排队玩过山车的孩子如今都已为人父母，游乐场所也从广州走到世界，然而南湖的湖水依旧碧绿如初。

第四章：栖水
Chapter 4: Waterfront Living

既是水之城，广州人自有一套驭水之道，或依水而造，取其水利之便；或引水绕村，享内航外航之便利，又自得小气候；或将湖泽化整为零，成护村周全的鱼塘；又或引江水入宅，活水成景，园内四季如春……广州人对水的理解，再透彻不过。

The local people have been versed at harnessing water, either using water for shipping and improving the microclimate in villages, or bringing water into courtyard for landscaping.

Chapter 4:Waterfront Living

Shawan Ancient Town

Year Built:
From Song to Qing Dynasties
Add: 10, Daxiangchong Lu, Shawan Town, Panyu District

Once a seaside area and now reclaimed land, the town is home to the densely distributed ancestral halls, old houses, streets and alleys. The village still keeps its layout from the Ming and Qing Dynasties while the town has some legacies from the seaside living in old days, house with walls made of oyster shells, which are typical for fishing

古海湾边上的这个镇，以水而名

沙湾古镇

曾为海湾之滨，后退为陆地。这里宗祠古屋、老街旧巷纵横密布，保存着完好的明清古村格局，镇上仍完好地保存了依海生活的旧痕——渔村常见的蚝壳屋。

年代 宋—清
地址 番禺区沙湾大巷涌路 10 号沙湾古镇

1	3	4
2	5	6

1—2.沙湾自珠江的沙洲上立村，历代繁衍生息，终成人口稠密的经济、文化重镇。
3.沙湾现在已渐离珠江江岸，唯有一堵堵蚝壳墙，留下了江与海的旧痕。
4—6.富庶之地、文明之乡，住所皆整洁，器具皆精美。

沙湾古镇位于番禺沙湾水道北岸，建于古海湾半月形沙滩之畔的"猪腰岗"，故得名"沙湾"。人们曾于沙湾清水井附近，挖出航海之物，且地下为白沙，故推测古镇曾在海湾之滨。沙湾古镇格局保存完好，保留了大量明、清时期与近代的古建筑，如具有岭南特色的镬耳屋、蚝壳墙，以及许多精巧的砖雕、木雕、石雕、灰塑等建筑艺术品。疏密有致的乡土树种、栩栩如生的壁画更添古镇怀旧情愫。

散落的风水池倒映着古朴的建筑细节，加上青砖铺地，宛如碧玉。植根于古镇的沙湾飘色、醒狮等民间传统文艺在此长盛不衰。

沙湾古镇中的留耕堂是番禺清代"四大宗祠"之首。全祠地势北高南低，为五开五进形制。其严谨的建筑布局、结构和典雅宏丽的装饰，展现了高超的古建筑艺术。

113

1	3
	4	5
2	

1.沙湾飘色，传统文艺长盛不衰。
2.何氏祖堂门前威仪。
3.留耕堂形制宏伟。
4.头门、仪门都相当精彩。
5.何氏是沙湾第一书香门第。

第四章 | 栖水
Chapter 4:Waterfront Living

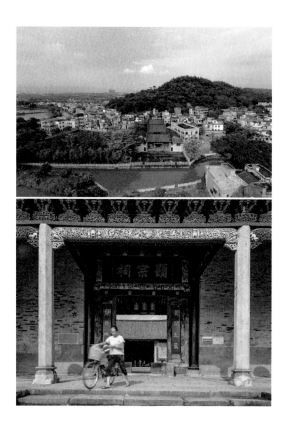

沙洲上的水文化典范

大岭村

具有完好的山、水、村、田的村落生态格局，也是珠江三角洲地区依水而居的传统村落的典型代表。

年代 宋—清
地址 番禺区石楼镇大岭村

大岭村位于番禺区石楼镇西北部，北面是广阔的耕地和鱼塘，中部是被誉为"七星岗"的七座山丘，砺江涌横贯东西。

1	3
2	4

1—2. 宋代兴村的大岭村，依山傍水、格局灵秀，屋宇华美、人丁兴旺。
3. 大岭村先人巧妙地将狮子洋之水引入村中，这条费人力挖出的玉带河，令全村出入交通便利、灌溉不假外求，同时水生风，形成宜居小气候。
4. 祠堂外有河水生风，屋宇对流畅达，三伏天仍是凉风习习。

117

大岭村原名菩山村,自宋朝开村至今逾800年,保存了大量具有岭南特色的文化遗产,其中最为著名的有南宋陈氏大宗祠"柳源堂"、明代显宗祠、清代龙津古桥,以及保留较完好的蚝壳墙等。

如诗"菩山环座后,玉带绕门前"所言,大岭村整体依山傍水,背靠菩山,三面河涌,贯穿全村的玉带河连接狮子洋,具有完好的山、水、村、田的村落生态格局,是保留较为完整的传统古村,也是珠江三角洲地区历史建筑的典型代表。

Daling Village

Year Built: From Song to Qing Dynasties
Add: Daling Village, Shilou Town, Panyu District

With its well-preserved ecological patterns of the village as represented by the mountains, water, village and farmland, Daling is a typical example of a traditional waterfront village in the Pearl River Delta.

1. 仪门外的落日。
2. 龙津桥上暮归的村民。
3. 风水塘、绕村运河、不远处的狮子洋……水,无处不在。

小桥入塘影，巷深人家绕
塱头古村

古村很好地保存了肌理完整的梳式布局，是研究岭南人居的完美样本。

年代 元—清
地址 花都区炭步镇塱头村

塱头古村于元朝立村，距今600多年历史，是迄今为止广东地区规模最大、保留最为完整的古村落之一，保存了很多明清时期广府特色的古建，气势恢宏的镬耳屋、风水塘前接连分布的书院是塱头古村精粹所在。

塱头古村为岭南广府地区富有特色的三间两廊梳式聚落，"红棉古树青云桥，小巷深处人家绕，书室栉比入塘影，渔樵耕读一梦遥"形容的正是这种布局。村落水系以半月塘为核心，分塱东、塱中和塱西三社，其中塱东社和塱中社相连，与塱西社以一条小河涌"深潭"相隔，体现了传统村落民居的风水布局理念、防御特色和生态特征。

Langtou Village

Year Built: From Yuan to Qing Dynasties
Add: Langtou Village, Tanbu Town, Huadu District

The Langtou Village is one of the largest ancient villages currently preserved in Guangdong. The buildings feature a comb-like layout with three halls and two courtyards. The water system centers on the Half-moon Pond, while the canals separate the village from outside on the east and west. This demonstrates the Fengshui-based layout of the houses in the traditional village, its defensive purpose and ecological characteristics.

第四章 | 栖水
Chapter 4:Waterfront Living

1		3			
2		4	5	6	7

1.塱头村现保存完整的明清年代青砖建筑有近两百座，祠堂、书室、书院错落有致。
2.塱头村的"塱"字，指的就是低洼的沼地，今天的塱头村仍保存着当年水系环抱的格局。
3.风水塘前，书院和祠堂并肩屹立，映衬出塱头村人尚文敬德之风。
4.书院和祠堂由深巷连接，形成精巧对称的梳式布局。
5.友兰公祠里的接旨亭记录下塱头村的辉煌岁月。
6.各家各院的神龛式样多有不同，砖雕技艺精湛。
7.谷诒书室虾公梁上的石雕非常精细华美，封檐板纹饰亦相当考究。

塱头村亦是广州人驭水成良居的一个出色范本：村人将白坭河水冲积出来的一片低洼地，开垦成既可防涝防盗、兼能养鱼饲鸭的连片水系。而古村南北向皆环山，夏季落山风带着水塘荷风拂南北向深巷，塱头村的乡居岁月，在酷夏里也是诗和画。

除了古树环绕、塘水相拥的大格局，古村现今保存完整的200座明清古建，如明代始建的渔隐公祠、留耕公祠，清代富商黄谷诒兴建的谷诒书室，都留有相当精彩的石雕、砖雕、木雕、灰塑。友兰公祠则保有明代皇室向塱头村黄氏颁授嘉奖的接旨亭。祠堂、书院比肩连片，炮楼、门楼气势雄伟，大格局恢宏、小细节精巧，依山环水的塱头村体现了广州人的居住哲思和审美。

123

第四章 | 栖水
Chapter 4:Waterfront Living

五邑商人的龙血宝地

聚龙村

聚龙村是一个有 120 多年历史的清代民居建筑群古村，迂回弯曲的珠江支流绕村而过，聚龙桥横跨其上，岸上的青砖大屋纵横交错、整齐有序，是广州老城区现存较完整的岭南建筑群落。

年代 清代
地址 荔湾区陇西直街 88 号

1	2	5
3	4	

1. 布列划一的聚龙古村，在城中隐于红尘，享时间的荫。
2. 既是商贾聚处，傍村运河自不可缺，河水接连花地河，通往大江大海。
3—5. 虽是南来北往、见多识广的富商巨贾云集之地，聚龙村仍留着东方的骨与魂。

Julong Village

Year Built: Qing Dynasty
Add: 88, Longxi Zhijie, Liwan District

Qulong Village, a 120-year-old village with typical Qing civil houses, is located by a tributary of the Pearl River, over which the Julong Bridge spans. The riverside brick mansions, with their well-organized layout, are known as well-preserved Lingnan building clusters in the old urban district.

聚龙村有百多年历史了，是广州老城区现存较完整的清代岭南民居建筑群落。当年台山富商邝氏家族依着珠江支流冲口涌边而建，冲口涌迂回弯曲，绕村而过，动土建房时岩底冒出朱红色的水，被称为"龙出血"，于是取名"聚龙"。

沿冲口涌两岸行，绿树掩映，跃龙桥架于涌上，一片生机盎然。聚龙村按井字平面构造，村中里巷纵横交错，整齐有序。建筑群前后分三排，前庭后街，坐北朝南的青砖大屋分布规整，结构布局相似。村前私塾出过不少商人和做官的人，也是聚龙村人杰地灵的一个佐证。

125

第四章│栖水
Chapter 4:Waterfront Living

将山水日月，藏在百年私宅
余荫山房

园内不见当年人，福荫尚留赠后辈。这是一座山与水、动与静、过去与未来融合的园林。

年代 清代
地址 番禺区南村镇东南角北大街

余荫山房是清代官人邬彬为纪念先祖、祈望子孙后代永泽福荫而建的园林，位列岭南四大名园之首，是广州仅存两座古典园林之一。

余荫山房结合岭南气候及文化风俗而建，以"藏而不露""缩龙成寸"的手法，将画馆楼台、轩榭山石亭桥尽纳于三亩之地，布成咫尺山林，园中有园，景中有景。选址背靠山峦，前有流溪，冬可御寒，夏可采光纳凉。园内以水造景，布局精巧，厅堂与楼阁、假山与流水动静结合。园中央筑有两池，架廊桥为纵轴，分隔成东西景区，曲水弯环、径随池转、廊引人随，有如长卷水墨画轴。所谓生态人居，便是这个样子。

Yuyin Mountain House

Year Built: Qing Dynasty
Add: Bei Dajie, Southeast Corner, Nancun Town, Panyu District

Ranking top among the four famous gardens in the Lingnan region and being one of the two most classic gardens in Guangzhou, the Yuyin Mountain House lies between the mountain and streams to stay warm in winter and cool in summer. With ingeniously designed layout and water features, the garden presents a charming view like a traditional ink painting.

1	3
2	4

1. 开轩面水，夏生凉风。
2. 水绕宅旁，便是风生水起。
3. 水榭，总是静心之所，也是开怀之地。
4. 水，使岭南庭院有了无穷的生趣与灵动。

1. 内有完备水系，外接紫坭水道，岭南的庭院，有大格局，也有大智慧。
2. 造水榭为观景，也成了水上的风景。
3—4. 岭南的庭院，总是不吝奢华富丽，处处用力雕琢。

Baomo Garden

Year Built: Qing Dynasty
Add: Zini Village, Shawan Town, Panyu District

Representative of the upright and honest officials, ancient Lingnan architecture, Lingnan garden art and the watery region of the Pearl River Delta, the Baomo Garden is an appealing antique garden and summer resort. It is also an art palace that one may explore tirelessly. The most amazing art work is the porcelain bas-relief version of 'Qingming Festival by the Riverside', which has been added to the Guinness Book of World Records.

以水为景，诗意栖居

宝墨园

一座园林艺术馆，同时也是一座藏品珍稀的艺术宫殿。

年代 始建于清代，1995年重建
地址 番禺区沙湾镇紫坭村

宝墨园原是包相府所在地，集清官文化、岭南古建筑、岭南园林艺术、珠三角水乡特色于一身，四时青翠，百卉丛开，是个美不胜收的大型仿古园林兼避暑胜地，更是一座文化宝物看不完的艺术宫殿。

全园水景，荔景湾、清平湖、宝墨湖与一千多米长河贯通，长流不息，若驾画舫轻舟，便仿似置身蓬瀛。正门白石仿古牌坊雄伟巍峨，园内治本堂后的"宝墨园"花岗岩石匾，是旧宝墨园内唯一真迹。园中陶塑、瓷塑、砖雕、灰塑、石刻、木雕等艺术精品琳琅满目，当中一幅瓷塑浮雕《清明上河图》已被列入上海大世界基尼斯之最。

第四章｜栖水
Chapter 4:Waterfront Living

珠水绕行江中岛，绿意盎然园中村

小洲村

有"广州南肺"之称的小洲村，保留了岭南水乡最后的小桥流水人家，见证过海上丝绸之路的商贸繁荣，惠泽着落地生根的文艺工作者。

年代 元—清
地址 海珠区小洲艺术村

这座由珠江潮水冲积而形成的小村落始建于元代，挨着万亩果林，有"广州南肺"之称。因珠水围绕而形成独特的村落肌理，村中水系和珠江水系相连通，河水随珠江涨落。

河涌蜿蜒迂回，民宅倚涌而建，街巷铺着麻石，静心细看，典型岭南风格的古建筑就在眼前。简氏大宗祠、天后宫、玉虚宫，灰筒瓦脊、砖雕灰塑精巧跳脱；今天难见真容的"蚝壳屋"也好好地保留在原地；村内几座晚清特色的商铺店号和"三间两廊"的特色民居，见证了海上丝绸之路的商贸繁华。

而让市民和游客乐此不疲的，是小洲村内的文化艺术工作者们改建的工作室、画廊、展馆、餐饮空间，他们的落地生根，使小洲村成了当代创意空间与传统水乡文化共融再生的试验场。

1	3	4
2		

1—2.参天树荫下，祠堂规整有序，造型各异的小桥横枕水上，一树一木、一屋一桥都在闲聊光阴。
3—4.昔日商贸繁华记在心头，今天先打好手上的牌，古村里时光悄悄溜走。

Xiaozhou Village

Year Built: From Yuan to Qing Dynasties
Add: Xiaozhou Art Village, Haizhu District

The small alluvial village neighboring the city's southern lung, the 10,000-mu Orchard, still preserves a typical view of the Lingnan watery region. While witnessing the old glory of the prosperous Maritime Silk Road, the village also benefits the cultural and art professionals settling down in the village.

第四章|栖水
Chapter 4:Waterfront Living

一渡到花岸，一渡到人家

云桂桥 | 利济桥 | 汇津桥 /

珠江两岸河涌密布，贯通海珠岛东西的马涌，如今仍存三座古桥，花岗岩、船形墩，其中云桂桥更是广州市区内现存最古老、保存最完好的石桥。河涌接入珠水，地灵人杰，古桥周边成了育才胜地。

云桂桥：明代，海珠区前进路晓港公园内
利济桥：清代，海珠区昌岗街
汇津桥：清代，海珠区马涌直街

| 1 | 3 | 1—2.云桂桥。
| 2 | 4 | 3.利济桥。
| | | 4.汇津桥。

珠江南端的海珠岛域，有一条东西贯通的河涌——马涌，马涌东端，正对着二沙岛，西端则通往珠江后航道。在长长的马涌河上，仍存有3座古桥。云桂桥、利济桥、汇津桥从东往西分布，至今涌边绿树依依，行人如鲫。

最东端的云桂桥在晓港公园里，是广州市区内现存最古老的、保存最完好的石桥。三桥主体皆为花岗岩，桥墩皆为船形，以减轻涌水对桥的冲击。

"云桂"二字，寓意"步云折桂"，体现了对读书人考取功名的美好祝愿，是明代清官何维柏于明嘉靖二十四年（1545）为方便学生出入来往所建。何维柏与著名清官海瑞齐名，返回家乡广州的何维柏全心执教，学生中举人、进士逾十人。

马涌的中段，利济桥、汇津桥之间，有另一处育才胜地——十香园，居氏兄弟带着一众弟子沿马涌写生，由此搭建出岭南画派的坚实基础。

Ancient Bridges

Add:

Yungui Bridge: Ming Dynasty, inside the Xiaogang Park, Qianjin Lu, Haizhu District
Liji Bridge: Qing Dynasty, Changgang Jie, Haizhu District
Huijin Bridge: Qing Dynasty, Machong Zhijie, Haizhu District

Both banks of the Pearl River were once densely distributed with small rivers and canals. The Machong Canal that winds through the Haizhu Island still has three old bridges featuring granite and ship-like piers. Among them, the Yungui Bridge is the earliest and the best-preserved stone bridge in downtown Guangzhou. The canal is connected to the Pearl River and inspires the intellectual development of the local people. The area around the old bridges has been well-known education community.

Chapter 4:Waterfront Living

龙津桥 | 石井桥 | 五眼桥

珠江水系庞杂，人们或借河流划分行政区域，过桥便是去到另一片新天地，比方五眼桥，一桥连通广州和佛山；又有桥扼交通之咽，一桥打通东西南北，比方石井桥，旧时广州至花县（今花都区）的驿道，靠它接连；更多时候，桥是水乡的肌理，像大岭村的龙津桥，从这一个家，到另一个家。

龙津桥：清代，荔湾区龙津西路
石井桥：明代，白云区石沙路石井桥
五眼桥：明代，荔湾区芳村石围塘五眼桥涌

建于清代的龙津桥，无声地展现着当时番禺大岭村大兴文化之风，红色砂砾岩建成的两孔拱桥，桥墩下是防止船只冲撞和较少水冲击的两个分水脊，栏板上有明代风格的卷草图案和暗八仙法器图案，桥侧还耸立着清光绪十年（1884）的六角形大魁阁塔。

芳村石围塘则有昔日省佛通衢的五眼桥，这座建于明代的古桥原名"通福桥"，横跨秀水河，明清时期，佛山南海的人们便经此桥进广州。而如今不甚起眼的五眼桥涌，当年可是广州大通内港的一部分，水面宽阔，常有阿拉伯、波斯等外国船只停泊，一片兴隆的景象。

而白云区的石井桥迄今也有170多年历史，清道光年间所建，是一座6个桥墩的梁式石桥。第二次鸦片战争时，广州人民组织团练武装抗击侵略者，此处石井乡的群众武装甚为活跃，现在石井桥的石栏板上还留有当时侵略军炮击的弹洞。有桥渡人，就有故事。

1	3
2	

1. 石井桥。
2. 龙津桥。
3. 五眼桥。

Two thousand years ago, Guangzhou people Ancient Bridges

Add:
Longjin Bridge: Qing Dynasty, Longjin Xilu, Liwan District
Shijing Bridge: Ming Dynasty, Shijing Bridge, Shisha Lu, Baiyun District
Wuyan Bridge: Ming Dynasty, Wuyan Bridge Canal, Shi Wei Tang, Fangcun, Liwan District

The ancient bridges in suburbs are of different views. The Longjin Bridge recreates the culture fever of the Daling Village in old days. The Wuyan Bridge, once the pivotal connection between Guangzhou and Foshan, now still offers convenient transportation for residents on both banks of the Xiushui River. The Shijing Bridge, a witness of the anti-invasion battle of the local people during the Second Opium War, still has the holes left from the invaders' bombards.

兴衰荣辱共一濠，
脉脉唱诵云水谣

"涌"只要在广州的路牌上出现，说明这里一定有过一条天然河。"濠"则不同，多为人工开凿以做护城之用。可见"东濠涌"这个名，便是——广州城东边的天然屏障被征用了。

从白云山流到珠江的东濠涌，肇始于宋的六脉渠。六脉渠由东濠涌接引通江，身兼航运、排污、泄洪、防火全职。明洪武三年（1370）疏通后更长更宽的东濠涌成了重要的防御工事，拥有了战略价值。

2	3
1	4

1.接麓湖之水为源，东濠涌的清波汨汨向南流去。
2—4.东濠涌所到之处，处处是碧水繁花、鸟鸣蝶舞。

第四章 | 栖水
Chapter 4:Waterfront Living

Donghao Canal

The Donghaochong extending from the Luhu Lake to the Pearl River is the only city moat left from old days. It was once acclaimed as "the CBD on water" and the No. 1 canal for shipping in the Qing Dynasty. By now the once polluted canal has been treated and restored to its original appearance and function, serving again for flood control and sewage drainage. It also makes a popular leisure destination for citizens.

兴衰荣辱共一濠，
脉脉唱诵云水谣

东濠涌

东濠涌北起麓湖，南衔珠江，是广州仅存的旧护城河。清朝货船往来，东濠涌跃居众濠之首，俨然水上CBD。历尽沧桑，如今东濠涌修旧如旧，重新发挥排污泄洪的职能，而当中的亲水平台和滨水公园，则成了市民夏日悠闲享乐的生活空间。

清朝，广州的商贸地位异军突起，货船往来的东濠涌跃居众濠之首，俨然水上CBD。风帘翠幕，烟柳画桥，一切皆可待价而沽。黄华塘有开中国机制钱币先河的广东造币厂，大东门的东平大押无言见证了多少家族的起落。

随着广州的陆路交通越来越发达，濠涌航运作用逐渐减弱直至消失。再加上久不清疏和居民侵占，水道日窄，洪涝频仍。随着工业化和城市化的进程加快，各类污水纷纷排入东濠涌，曾经为广州人提供生活用水的地方竟成了周边居民连窗都不敢开的臭水沟。

从2009年开始，广州市全力推进大型治水工程。东濠涌因其穿过环市路、东风路、中山路三条中心城区主干道而成为最重要也"最难啃的骨头"。先截污再雨污分流，然后利用潮汐把珠江水引入现有河涌，形成"东濠涌—珠江—东濠涌—珠江"的水流大循环。通过混合絮凝、气浮过滤、紫外线光催化氧化等流程，对涌水进行净化并保证达到4类以上水标准。

水清了，岸也跟着绿了，不仅要看得见的绿，更要摸得着的绿，沿岸的滨水公园有夏日傍晚最悠闲的徜徉。东濠涌里挖走的是淤泥，挖出的是全方位多角度的广府文化。

依照"修旧如旧"的原则，恢复了沿线的越秀桥、小东门桥、筑横沙桥、东华桥等6座历史名桥原貌，鲁迅故居等名人故居随着东濠涌变美再次被擦亮。

以河涌为主题的东濠涌博物馆原本是涌旁的一栋民居，曾经它只是忍受东濠涌的千门万户中的一个，现在它成了留守的唯一建筑。广州人跟东濠涌的居住距离看似远了，但他们跟东濠涌的生活距离却更近了。

1	3	4
2		

1.六脉渠仅余的东濠涌，经过大力整治后，沿涌皆有景观。
2.涌上一架飞虹行车，桥下一湾清水奔涌。
3—4.鲁迅便曾在这楼里，倚着东濠涌，听过珠江的潮歌。

第四章|栖水
Chapter 4:Waterfront Living

1	
2	3

1—3.将荔枝湾涌恢复成飞廊架涌水、清流绕人家的岭南水乡景观，将环线众多的人文景观和自然美景有机串联，昔日荔枝湾风光，幻变成今日处处可入画的风景。

一湾江水见新绿，
两岸建筑出网红

荔枝湾涌

2000多年历史的荔枝湾涌，在老广心中从来不可取代，走一趟改造活化后的荔枝湾涌，忆当年羊城八景之"荔湾渔唱"，岭南广府文化精华尽收眼底。

当年汉高祖刘邦派遣使者陆贾到广州劝降南越王赵佗，陆贾在其驻地外溪水旁种了荔枝、莲藕，"荔枝湾"因而得名。明代，渔民早出打鱼，晚归停靠，于是有了羊城八景"荔湾渔唱"。而今老广口中的"荔枝湾"，则是依着荔湾湖公园一带的清末新荔枝湾涌。

物换星移，近代荔枝湾涌水质堪忧，一度被掩埋覆盖。而因广州亚运会，荔枝湾涌迎来一次华丽变身，人们记忆里"一湾江水绿，两岸荔枝红"的荔枝湾迎来新生。河涌被活化，水网重现，两岸绿树红花惹人醉，沿岸的名伶故居等珍贵古建也展露了真颜，岭南水乡的景观不再停留于诗情画意，而是成了市民和游客看得见摸得着的风景。

Lychee Bay

The over 2,000-year-old Lizhiwan Canal is indispensable in the local people's life with its water network, lychee, freshwater products and songs of fishermen. At its prime time in the Ming Dynasty, it was included in the Eight Scenes of Guangzhou. Nowadays the revitalization of the canal unveils the precious historical buildings on both banks.

岭南两大文化标志，汇于涌畔

粤剧艺术博物馆

荔湾涌畔曾是广州粤剧活动的重要聚集地，今岸线空间除了一系列珍贵历史建筑，还有一座以传承粤剧为己任的岭南传统园林就地而起，岭南两大文化标志实现了一场跨越时空的对话。

年代 2012
地址 荔湾区恩宁路127号

荔枝湾涌改造项目以荔湾湖为依靠，从自然景观和人文风情两方面出发，联系周边自然水林资源，恢复河涌，调水补水，改善水质。同时，原藏于街坊内部的陈廉伯公馆、文塔、梁家祠等历史建筑错落分布在涌边，与邻岸民居建筑相呼应，融合岭南园林元素的廊、亭、桥、叠石流水等成为公共空间。堤岸通过标高变化形成多层次的岸线空间，加上若干埠头，游船游人如梭，河涌便真正活了起来。

这光景让人忆起繁盛时期的荔枝湾涌，沿岸众多粤剧私伙局，园林、茶楼皆是演艺空间，粤剧历史建筑八和会馆也隐身于荔枝湾涌附近，"南国红豆"的美誉就这样随着珠水传遍世界。今天为了保护和传承广东省唯一世界非物质文化遗产——粤剧艺术，在荔湾涌沿岸新建了仿岭南古典园林建筑群——粤剧艺术博物馆，岭南两大文化标志"岭南园林"与"粤剧艺术"实现了一场跨越时空的对话，园林式布局与河涌景观及城市原有肌理和谐相融，馆内汇集岭南传统工艺"三雕两塑"之精品，公众的参与更使这座博物馆成为荔湾涌上的又一地标。

| 1 | 3 | 4 |
| 2 | | 5 |

1—2.荔枝湾涌改造项目中的瑰宝——粤剧艺术博物馆，将岭南水乡之风貌与园林之美完美地呈现于人前。
3—5.粤剧艺术博物馆木雕、砖雕、石雕、灰塑、陶塑、嵌瓷等等工艺，无一不显示出广东民间艺术的超凡技艺。

Cantonese Opera Art Museum

Year Built: 2012
Add: 127, En'ning Lu, Liwan District

The Lizhiwan Canal waterfront and the old downtown of Xiguan used to be the city's popular destination for Cantonese Opera events and celebrities. The garden-style layout and canal landscaping well merge with the existing urban fabrics. The public participation makes the museum another landmark by the Lizhiwan Canal.

第四章 | 栖水
Chapter 4:Waterfront Living

1	4
2	3	5

1.整治后的猎德涌，为钢筋铁骨的珠江新城增添了杨柳依依的柔媚。
2—3.祠堂、楼阁、老树、依旧年轻的猎德涌……时光的痕迹，无处不在。
4.猎德涌在整治时，留有藏龙塘，每年端午，此涌便再现千舟竞渡的热闹场面。
5.传统与现代，便在一涌间悄然融合。

猎德而得德，
千年河涌的延续与再生

猎德涌

流经珠江新城 CBD 的猎德涌，是广州市四大重要水脉之一，也是近年城市中心区景观建设的重要一环，经过治理后的猎德涌地区，现代建筑与传统村落的完美融合，使广州的河涌文化得以再现眼前。

猎德涌，旧时又称"猎水河"，隔着900多年的苍茫时光沿水路回溯，在北宋之前，猎水只不过是广州古城东南外的一条郊野水系。

直至北宋元丰年间（1078—1085），猎德开村始祖李铨举家从陇西南迁，来到猎德定居，开垦芜田为耕地，疏浚通水，渐成村庄，彼时猎德涌便成为村庄灌溉、供水的主要渠道。

猎德在宋时能位列广州八大重镇之一，究其缘由，因猎德涌与珠江交汇的猎德码头，有极为便利的水运条件。而后的数百年间，猎德码头发展成为中西商品、文化的交会点。两岸的村落也逐年因村繁荣兴旺。

旧时猎德涌清波碧水，居食可饮，除了供沿线居民泛水舟运、生活给水外，还是宗祠文化、端午赛龙舟活动、拜埠头祈福活动、中秋烧花塔等重大民俗活动的载体。

几百年过去，猎德涌两岸虽已是高楼拔地起的现代商贸重地，但肌理仍是几百年来织就的水乡格局，埠头、码头连着河涌与陆地，两岸青石板路边的民居、祠堂、私塾临河而立，古树葱郁，人口稠密。

今天的猎德涌，是天河区唯一流经珠江新城 CBD 的河涌，它从华工西湖流经天河商务区，再到珠江新城猎德村，最终汇入珠江前航道、奔涌向海。

Liede Canal

The Liede Canal, which winds through today's prestigious Zhujiang New Town CBD, is one of the city's four major canals. It functions as the flood control and forms an important part of the landscaping in the city center. The harmonious integration of the modern buildings and traditional village in Liede promotes the revival of the canal-based traditional culture.

第四章｜栖水
Chapter 4:Waterfront Living

1	4
2	3

1."猎德"，源自西汉著名思想家扬雄在《法言义疏·学行》中所言："耕道而得道，猎德而得德。"猎德的意思是追求完美的道德的意思，因此，村赐水名而水使村兴。

2—4.猎德涌是流经珠江新城CBD内唯一的河涌，水景丰富了广州新中轴线景观空间，和相隔不远的广州歌剧院、广东省博物馆新馆、电视观光塔及东西双子塔相映生辉，历史与现代在潺潺的流水中沉淀交融，民俗文化与当代生活互为平衡糅合。

　　随着城市化不断推进，猎德涌曾一度与广州其他河涌一样，水体污染严重，两岸居住条件恶化。最终，广州以大手笔，分三次对猎德涌进行截污整改，今天，这条CBD里的景观河涌重焕生机，生态系统实现良性循环，清水长流、绿树成荫。

　　无论是文教区的亲水小溪、商业区的繁华水岸，还是拆迁后的猎德村民依旧制重塑的古建、宗祠、庙宇……都在展现着猎德深厚的南国水乡风情，龙舟竞渡、拜埠头等活动在这里延续下去，传统与现代在川流不息的碧波中沉淀与交融，民俗文化与当代生活在这块水滋养的土地上互相平衡、补充。

　　正如猎德村复建的祠堂里的这副对联所书："河光不随流水去，天风直送海涛来"，横批"德泽天河"，这被泽被的福地，果然猎德而得德。

147

作者名录
Author List

本书收录了众多机构和摄影师提供的精彩的图片，在此表示感谢。图片的版权归属于图片的提供者。

This book contains wonderful pictures provided by many institutions and photographers, to whom we express our thanks. The copyright of the images belongs to the providers of the images.

大写岭南：
P4 P5 P6 P7 P8 P9 P10 P11 P12 P15-1 P16-2 P17 P18
P19-2 P20-6 P21 P31 P32 P35 P36 P37 P38 P39 P40 P41 P42
P43 P44 P45 P46 P47 P48 P49 P50 P51 P52 P53 P54 P55 P56
P57 P58 P59 P61 P62 P63-1 P64 P65 P66 P68 P69 P70 P71
P72 P73 P74 P75 P76 P77 P78 P81 P82 P84-3 P85 P86 P87 P88
P89 P90 P93 P94 P95 P96 P97 P98 P99 P100 P101 P102
P103 P104 P105 P109 P110 P111 P112 P113 P114 P115 P116
P117 P118 P119 P120 P121 P122 P125 P126 P127 P128 P129
P130 P131 P132 P133 P134 P135 P136 P137 P138 P139 P140
P141 P142 P143 P144

大写岭南（插画）：
P4 P5 P6 P7 P8 P9 P10 P29 P30 P33 P34 P71 P72 P91 P92
P107 P108

中国图片库：
P67 P79-2 P80 P82-5

测绘局：
P1

广州市城市规划设计有限公司：
P13 P15-2 P16-1 P19-3 P59-2 P79-1 P97-2
P106 P123 P124

陈文杰：
P83 P84-4

胡建军：
P20-6

黄得珊：
P19-1

罗宜威：
P2 P60 P63-2

龙建平：
P19-4

杨艺：
P14-5 P16-2 P20-5

曾贵鸿：
P21-8

摄影作者：
陈文杰 陈欣 耳东尘 胡建军 黄得珊
黄庆衡 李波 李卫东 梁洁红 梁宇星
刘朝宽 刘晓明 龙建平 罗宜威 吕凤霄
孟俊锋 欧阳永康 彭庆凯 饶国兴 史丹妮
孙兰 覃光辉 王玉龙 温建红 吴术球
叶秉新 张秀珍 邹庆辉

文字作者：史丹妮 郑宇 孙海刚 崔玛莉
张远 莫尔多姿

插画作者：高毅 龙志放 陈丁财

翻译：梁玲